Elements of African Bioethics in a Western Frame

Godfrey B. Tangwa

Langaa Research & Publishing CIG
Mankon, Bamenda

Publisher:
Langaa RPCIG
Langaa Research & Publishing Common Initiative Group
P.O. Box 902 Mankon
Bamenda
North West Region
Cameroon
Langaagrp@gmail.com
www.langaa-rpcig.net

Distributed in and outside N. America by African Books Collective
orders@africanbookscollective.com
www.africanbookcollective.com

ISBN: 9956-578-15-0

Table of Contents

Introduction

About 1996, Peter Singer, former President of the International Association of Bioethics (IAB) and Editor of *Bioethics*, journal of the said association, approached me with a request to write a paper on the possibility of non-Western bioethics. This invitation came out from the increasing realization that no single voice, nor perspective nor culture can claim to represent the whole of bioethics. I responded to this invitation by attempting to open a small window on elements of traditional African bioethics drawn from my own natal culture, Nso' culture of the Bamenda region of Cameroon, a predominantly oral culture, like most other African cultures. My effort resulted in the article "Bioethics: An African Perspective" which was published in July 1996 in *Bioethics*, Vol. 10, No. 3, pp. 183-200. That paper forms Chapter Two of the present book.

The explosion of bioethics in the contemporary world is largely due to problems and concerns connected with Western biotechnology and with developments in Western biomedicine. This easily leaves the impression that bioethics is not only a purely Western affair but that it is also of quite recent origin. But if we consider bioethics as the ethics of the biological world, as human concerns over problems connected with life and living in general, including vegetal, animal and human life, then it is clear that bioethics could not have come into existence only with the Western technological revolution and cannot be a purely Western affair. There are important bioethical perspectives, insights and values in non-Western contexts, which might illuminate and contribute to the solution of contemporary bioethical problems, dilemmas and quandaries, even if these arise mainly in connection with Western biotechnology and biomedicine.

There are definitely African perspectives on bioethics. How could there not be? For millenia, Africans have lived on the African continent, in close contact with the diversities of nature: floral, faunal and human; and in so doing they have developed cultures, values, attitudes and perspevtives to the problems, ethical and otherwise, that have arisen from the existential pressures of their situation. The problem, however, is that such values and perspectives do not necessarily form coherent ethical theories. Theory-making is a second order activity. The elements of African bioethics are to be found in its cultural values, traditions, customs and practices. These are research-able, highlight-able and usable by those who would.

The bioethical problems of our current global existential situation are such that all possible solutions, no matter their provenance, ought to be tried. Western thought patterns and values, practices, and ways of doing things have gained the status of global paradigms, largely thanks to colonization and proselytization by Westerners of non-Westerners. Western culture has far too loud a voice combined with deaf ears in contemporary ethical discourse. But it should never be forgotten that other cultures have their own word to say and that alternative values, ways of thinking and practices exist, and attempt should always be made to bring these out and to highlight them, if they could possibly contribute to the satisfactory solution of a global problem.

Power and influence may often be serious impediments to perception or acceptance of simple hard truths and particularly of ethical rightness or appropriateness. That is why, for example, it is much easier to reach global concensus that people like Charles Taylor or Omar El Bachir are self-evidently war criminals than that Tony Blair or George Bush should be considered in similar light for having cool-headedly, rationally and calculatively gone to war in Iraq on a false claim, knowing fully well that it would result in the death of thousands of innocent human beings. The problem of double standards is one of the most important problems of contemporary bioethics. In the end, power, influence and prescribed political correctness, rather than rationality or ethical correctness, are what would seem to rule the world. The ethical conscience should not be sleeping soundly in the face of such situation.

In this book, I have brought together various papers on bioethical issues and problems, written at different times, some previously published, each of which attempts to bring out some African elements, perspective or concern, in a way and manner that would be comprehensible within Western culture and perspective. The African narrative style predominates through these essays but their framing conforms, more or less, to the Western paradigm for presenting academic issues.

The first chapter is an interview I had with a students' magazine of the University of Groningen, The Netherlands, in the course of which I give some biographical background of myself and discuss some of my current bioethical preoccupations and concerns. The second chapter is my paper on an African perspective on bioethics, substantially as published in *Bioethics* in 1996. The third chapter is a paper on Birth and Death developed from a presentation I made during the Third World Congress of Bioethics in San Francisco, USA, in 1996. The fourth chapter is a paper on the abortion debate, previously published in a local magazine of the Ministry of Public Health of Cameroon. The fifth chapter is a paper on bioethics and sustainable development, first written on the invitation of the World Business Council for Sustainable Development. The sixth chapter was first written at the invitation of Professor Kwasi Wiredu for *A Companion to African Philosophy* of which he is the editor. The seventh chapter on rights and rationing in healthcare was first presented at a seminar at the University of Bristol, the UK, organized by Alasdair MacIntyre in 1999. The eighth chapter on morality and culture has previously been published in the Turkish journal of Medical Ethics, Law and History in 2004 and in *Bioethics in a Small World,* edited by F. Thiele and R. E. Ashcroft, in 2005. The nineth chapter was first made as a presentation during the Fifth International Symposium on Sexual Mutilations at Oxford University, England, August 4-7, 1998, and subsequently published in *Male and Female Circumcision: Medical, Legal, and Ethical Considerations in Pediatric Practice,* edited by George Denniston et al. in 1999. The tenth chapter on Gender, Feminism and Motherhood, which probably unknowingly breaches some Western taboos, has been presented in several conferences and symposia and is published here for the first time. The eleventh chapter has also been presented at several conferences and is being published here for the first time. The twelveth chapter is essentially a short pep story first told at the Fourth International Tsukuba Bioethics Roundtable, University of Tsukuba, Japan, in 1998, and subsequently in the *Newsletter of the International Association of Bioethics (IAB)*. The thirteenth and final chapter developed from a presentation delivered on 21 October 2008, at the German Presidency, Schloss Bellevue, on the invitation of the German President, Horst Kohler.

Some ideas, metaphors or arguments recur over several of the chapters of this book, hopefully giving to the entire work a semblance to an African folk tale, with its usually characteristic refrains and repeatitiveness.

Finally, it is my fervent hope that this book would inspire African scholars interested in the field of bioethics to diligently research into African culture and traditions for ideas and values that could be salvaged from their recession into oblivion and given analytic viability in global bioethics discourse.

Godfrey B. Tangwa
Yaounde, Cameroon
15 August 2010

CHAPTER ONE
INTERVIEW WITH PROF. DR. G.B. TANGWA

[This interview, conducted by Geert van der Velde in April 2006, for the Students Magazine of the University of Groningen, The Netherlands, was a follow-up to a symposium lecture delivered by me, at the invitation of the Medical Students of Groningen University, in December 2005, under the theme: „Western Research into Non-Western Diseases"]

It is interesting to see how philosophers of different cultures and backgrounds challenge the same problems in a different way. Professor Dr. Tangwa lives in Cameroon and finds himself in a country which fights against famine, AIDS, exploitation by western countries and companies, to mention but a few of the recurring themes in his everyday life. As a philosopher Dr. Tangwa challenges these problems, and specifically Cameroon's, and while doing so he draws upon African culture - his own Nso heritage in particular - to find answers for many of these urgent questions.

Question: In many of your articles you start from problems that are directly related to Africa or Africa's fate. You also draw on African (Nso) culture in your attempts to come up with solutions to these problems. Could you tell us a little about your own background and the culture you draw so much inspiration from?

Tangwa: I was born into an extended family in a large African compound called 'Lum', in the village of Ndzenshwai-Shisong, in the Fondom (Kingdom) of Nso', in the Northwestern (Bamenda) highlands of Cameroon. My biological parents were among the Christian converts of the lineage, so I was baptised at birth and had the privilege in my early upbringing of both traditional (pagan, if you like) and Christian influences. The first Christian (Catholic) mission in the whole of the Bamenda region had been established in Shisong in 1912 by two German priests, Lennartz and Emontz, of the Society of the Sacred Heart of Jesus, sent by their mother house, Sittard in Holland. Ndzenswhai was one of about six quarters or sub-villages of Shisong, being the most densely populated and the only one with many traditional families. The other quarters of Shisong were inhabited mostly by Christian converts, many of whom had migrated from other villages of the Nso' Fondom, to live close to the Church and sometimes to escape from the religious persecution of their 'pagan' kith and kin.

Ndzenswhai is nicely separated from the rest of the other quarters of Shisong by the River Shwai. Around the early nineteen hundreds, when the Fon (King) of Nso', Nga'-Bihfon I, gave Shisong to the Catholic missionaries for settlement, Nzdenswhai comprised about half a dozen family lineages (Kilam, Wang, Kigom, Lum, Mbiim and Lavkishwang); the last-mentioned of which is no longer in existence today, having completely relocated to another part of the Fondom, while all the others have suffered a diminution in population strength, physical appurtenances and general visibility, owing to rural-urban migration, erosion of the traditional fabric of life, general impoverishment and the devastating effects of modern epidemics.

I am basically a villager in my dispositions, attitudes and innate expectations/reactions and, in spite of having widely travelled the world, Ndzenshwai-Shisong remains the only place on planet Earth where I feel completely at home and at peace with myself. And yet, as of today (2006), Ndzenshwai has neither electricity nor a paved road, our collective efforts, entreaties and expectations, in this regard, having been consistently frustrated by the governmental people of Cameroon. We do, however, have a small pipe-borne water scheme which resulted from an initiative of one of my European friends, Sue Willdig, who helped us arrange initial funding for the project from the German NGO, *Missereor*.

The pioneer Catholic missionaries established two Western-style primary schools in Shisong, one for boys and the other for girls. I went to primary school at the age of 3 and, upon completion, decided to follow the path of the Catholic priesthood, following the encouragement particularly of many Reverend sisters (Shisong also had a convent) and also that of my parents. At the seminary, I suffered a crisis of conscience which convinced me that I was more of a pagan than a Christian in the depths of my heart, and (doctrinally) a potential heretic, as a future priest. In spite of being rated a brilliant student and exemplary candidate for the priesthood, I voluntarily decided to quit the path to the priesthood, to the disappointment, if not consternation, of many including my family, and thereafter pursued secular studies.

Nso' culture is a very communal culture, organized on the visual model of concentric circles, representing hierarchical centres of traditional authority. The smallest of these circles is the nuclear family in its western acceptation, which in the African context is properly called a 'house-hold', comprising a man, his wife or wives and their offspring. Next, comes the immediate lineage, headed by a lineage head (Taala'/ Fai or Sheey), usually designated by the Fon (King) whenever the position becomes vacant. The lineage head has very important responsibilities for the material, social and spiritual well being of each and every member of the lineage and for the growth and expansion of the lineage. He is usually a custodian, priest and healer in one. Next, comes the extended lineage, usually headed by a 'Shuu-Fai' (councillor to the King) to whom allied lineage heads pay homage and allegiance and from whom they draw inspiration, advice and mutual support. Lastly, comes the King, the chief high priest of Nso', titular owner and chief custodian of all land in the kingdom, final judge and arbiter of all disputes.

Q: In the article: "The HIV/AIDS pandemic, African traditional values and the research for a vaccine in Africa' you discuss the various disadvantages of a 'Western approach' and, inspired by the culture you just told us abou,t you emphasize the importance of African communitarian values. Could you elaborate or give us a good illustration of a more 'African' approach to the problem of HIV/AIDS?

Tangwa: A 'more African' approach to the problem of HIV/AIDS would first recognize it as an epidemic threatening the very survival of the entire community and therefore requiring the urgent mobilization of all of the community's resources in fighting against it. HIV/AIDS is a global epidemic requiring the mobilization of global resources in fighting against it. This has not been the case, as the best resources of the developed industrialized communities of the world, who are less at risk and less affected, are lavished on enhancement medical care and other luxury medical researches, while medicines which can mitigate the effects of HIV and delay its lethal *terminus ad quem*, widely available in the developed world, are mostly inaccessible or unaffordable in the developing world. In the traditional African setting neither social nor economic status was a barrier to accessing treatment if one was ill.

Q. Why do you think neither social nor economic status was ever a barrier to accessing treatment in traditional Africa?

Tangwa: Because money was not involved; the art of healing was not a commercial activity or occupation one could practice for the purpose of earning a living. Consultation of healers and access to treatment were quite free, except in so far as the patient might be asked, in certain cases, to provide some of the common ingredients necessary for preparing his/her medicine.

Q. What reason(s) do you think there is/are for the (relative) inaccessibility and stratification of (medical) treatment in Western societies?

Tangwa: The main reason, I believe, has to do with the marriage between medicine and the market, the evolution of medicine into a purely commercial and highly lucrative activity in which investments understandably yield enormous profits.

Q: You state that the Western economic idea and practice (the more desperately you need a product or service, the more you are required to pay for it under the so-called law of demand and supply) has unfortunately become globally accepted. 'Wouldn't you say that this kind of calculating, distracted from local commitments and values, this 'Western approach', is inherent to globalisation?

Tangwa: Yes, in its economic dimension; and many people would argue that globalization is nothing more and nothing less than an economic process whereby the industrialized Western world has taken command of the world's economy and is economically colonizing the entire globe. Personally, I also look optimistically on globalization as an opportunity for the dominant/domineering Western world to lend an attentive ear to the other cultures of the world, to appreciate and learn from their cultural values, to relinquish the initiative of their autonomy and self determination, so that the emerging global culture will be truly a culture that has benefited from the best values of all the cultures of the world; for no human culture is perfect and none so poor that it has not discovered human values that have eluded the others. Globalization does have other dimensions, including a prescriptive dimension and, if the lessons of postmodernism are correctly learned, the industrialized Western world would not strive to impose its economic determinism and 'might is right' operational philosophy on all other societies; postmodernism should go hand in hand with postcolonialism. Globalization should not just mean Westernization, let alone Americanization, where world policy is set with strident declarations from the White House.

A globalized economy does not mean one which follows the dictates of some supposedly omniscient controllers in Washington, London, Paris, Bretton Woods etc., but rather one in which all the inter-dependent communities of the world freely exchange their produce in a common market, regulated by unmanipulated market forces. If globalization is to mean expanding and imposing the highly exploitative Western economic system to other parts of the world, then I am opposed to it.

Q. Wouldn't you say that a kind of criticism of irresponsibility on a global level would be more effective to address the problems of a globalizing world than an appeal to local identities or values?

Tangwa: Criticism at a global level is important, but it needs to evolve from recognition of local identities and the fact that the dominant Western identity is also a local identity imposing itself globally.

Q: You call yourself a 'cultural pluralist' and a 'moral universalist'. How do you reconcile those two positions?

Tangwa: Culture is the way of life of a group of people, underpinned by adaptation to a particular environment, a shared worldview, ideas, values, historical experiences, attitudes, expectations, practices, etc. As such, cultures are forms of life analogous to biological species. Cultures *qua* culture are a *datum* of the social nature of humans. Cultural pluralism is inescapable, because there is no reason to think, let alone propose, that any particular culture as a culture should not exist.

Morality, on the other hand, is concerned about right/wrong, good/bad in human actions/behaviour, including actions/behaviour that are cultural practices. What is right/wrong, good/bad in the moral sense could not differ from place to place or from time to time, at the moment of perception. Morality not only is necessarily prescriptive, it is necessarily universal in its ambitions, although, like all other things human, it is subject to human epistemological and other limitations.

Q. But wouldn't you say that morality is concerned with a way of thinking about right or wrong and/or with moral feelings about the good and the bad in human actions/behaviour. And are these feelings not connected with a way of life of a group or people, their ideas, worldview, values, etc? How can we, in an epistemological sense, possibly make the distinction between what we experience as right/wrong or what we think is good/bad and what is right/wrong or is good/bad? And if not, is morality - connected in this way with the experience of what is right/wrong, so with the worldview, values and ideas of a group of people - not an expression of a people instead of all people?

Tangwa: It is an expression of a people in all their epistemological limitations, necessarily aimed or intended at having validity and applicability for all the people.

Q: You believe that 'disinterested reflective deliberation' can narrow (or close?) the gap(s) in moral diversity. How do you propose we come to a 'disinterested reflective' point of view through deliberation?

Tangwa: The main impediments to disinterested reflective deliberation, in my opinion, are egoism and the belief that might is right. Egocentrism is a psychological datum which need not issue in egoism. The impediments to disinterested reflective equilibrium can be subverted by an emphasis on the absolute moral equality of all humans, on the fact that truth is never like a hammer and on the disastrous historical errors of the mighty in all epochs of human history. It also needs emphasizing that self-interest on which some powerful cultures purport to found their moralities could not possibly be the foundation of morality, inasmuch as morality comes in at all only where the self is side by side or up against other selves, implying the impossibility of proceeding like a solipsist, like an absolute egoist.

Q: What kind of practices and institutions do you think are compatible/useful with this ideal?

Tangwa: For practices, democracy and, for institutions, a United Nations Organization, of equal and permanent members, none with veto powers, not given only to manipulative power politics, but to deliberations culminating in consensus, with global well-being as its main preoccupation.

Q: Could this valuation of a disinterested reflective deliberation and a disinterested reflective point of view be an ideal or an expression of a certain moral engagement which is itself culturally dependent? Isn't it possible, for instance, to proceed from a universally accepted plurality of (radically) different views?

Tangwa: It is possible to proceed from a universally accepted plurality of different views, provided the ineradicable limitations of each and every one of those views are first recognized and fanatical self-righteousness combated.

Q: You fight for a more cooperative and sympathetic world. You state for instance that 'it is, first and foremost, the time to mobilize all available resources in the interest of those helplessly in need.' What does this mean for the individual, practically speaking?

Tangwa: There is a moral imperative here, deducible from the very idea that we are human beings. From the point of view of the individual, there are two possible areas of action: individual philanthropy and collective action in cooperation with others aimed at influencing government policies and actions.

Q: What kind of role do you see for the ethicist? Do his or her opinions have an influence?

Tangwa: Ethics is prescriptive, and the role of the ethicist is persuasive and exhortatory, aimed at achieving results by convincing through rational arguments. The ethicist may not achieve results as fast as those who use other methods, such as force or lobbying, but the results of the ethicist, when achieved, are more influential, firmer and longer lasting.

CHAPTER TWO

BIOETHICS: AN AFRICAN PERSPECTIVE[1]

[This chapter was first published in the journal of the International Association of Bioethics (IAB), BIOETHICS, Volume Ten, Number Three, July 1996. It was written at the instigation of Peter Singer, editor of the journal, for a special issue, under the theme: The Prospect for 'Non-Western' Bioethics.]

PRELIMINARY REMARKS

In attempting to say something about an African perspective of Bioethics, one is haunted and daunted by the historically notorious question: "Is there such a thing as African philosophy?" on which western-trained African philosophers squandered nearly two decades in debates which often produced more heat than light.[2] Given certain **assumptions**, the presumption that there is an African Bioethics

---

[1] I wish to thank the German Alexander von Humboldt-Stiftung for a fellowship award, tenable in Germany from October 1994 to June 1996, which gave me the opportunity of having first-hand experience of certain aspects of western culture that I would otherwise not have had. I am grateful to the Archives Section of the Library of the London School of Economics and Political Science (LSE) for permission to consult P.M. Kaberry's unpublished manuscripts and Fieldnotes and to E.M. Chilver for a stimulating discussion of these, started at Oxford on 21 November 1995, and continued by mail. I am also grateful to Wole Ogundele for allowing me to see the unpublished transcripts of his conversations with Ulli Beier. My use of the German version "Kamerun" in preference to the English "Cameroon" and the French "Cameroun" throughout this essay is quite deliberate and prescriptive for reasons fully expressed in my (as yet) unpublished paper : "Colonialism and Linguistic Dilemmas in Africa : Cameroon as a Paradigm".
[2] A good glimpse of the more formal content of these debates can be found in the following sources, among many others: Paulin Hountondji, *African Philosophy: Myth and Reality*, London, Hutchinson, 1983; Richard Wright, *African Philosophy: An Introduction*, Washington, D.C., University Press, 1979; H. Odera Oruka, *Sage Philosophy: Indigenous Thinkers and Modern Debate on African Philosophy*, Leiden, E.J. Brill, 1990. See also my two-part article: "African Philosophy: Appraisal of a Recurrent Problematic" in *COGITO*: Summer 1992 and Winter 1992. The informal aspects of the controversy were, however, equally significant.

worthy of the serious consideration of thinkers, especially ethical enquirers, world-wide, may be daunting to substantiate. Such putative assumptions might, for instance, be that Bioethics is something only to be encountered in books and journals and discussed during lectures, conferences, symposia, etc. Given such and similar assumptions, the task of showing that traditional African Bioethics exists and is worthy of consideration by non-Africans may be as difficult as the task of demonstrating that traditional African Religion, for instance, exists and is worthy of similar consideration. In the latter case, the incredulous reaction might be:

> "Wonderful! Quite fascinating!! Could you kindly describe this African Religion for us, systematically and coherently? Who was its founder? What is its Holy Book or sacred script? Who is the head of the faith and who are its most authoritative theologians? What is the approximate membership of the faithful? How much would an average preacher realise on average from the "Collection" during a typical service? What architectural style is used in building its churches? What do the liturgical vestments look like and what are the main features of the liturgy itself?"

To admit that none of such questions can be answered but continue to insist that there is such a thing as African Religion worthy of the careful attention of those who inevitably think of Religion in the above terms and categories, is to set oneself a task that requires great courage even if, in the case of Religion, a very promising point of departure might be to draw attention to the palpable fact that religion as conceived above has been the source and origin of some of the most irreligious and horrendous experiences of humankind.

WESTERN CULTURE AND AFRICAN CULTURE
By virtue of its technological advancement, western culture is today the dominant culture of the world. This domination has been greatly assisted by the intrusion, in the form of imperialism, colonisation and neo-colonialism, of the western world in non-western worlds. The scientific-cum-technological success of the western world has, furthermore, made it to consider itself and to be generally accepted as an infallible oracle on all other spheres and all other matters. The western point of view and western philosophies and practices are

everywhere loudly propagated and widely disseminated with continuing arrogance. Because of this, western culture has developed a high degree of immunity and imperviousness against influences from other cultures except in areas where these are inevitable corollaries of its exploitative activities.

The dominant philosophies and ideologies of the western world, especially as practised though not always as propagated, accept and promote the doctrine that "might is right". This, taken together with the belief that "knowledge is power" leads to a pervasive desire to know everything, to co-ordinate everything, to unify, to harmonise, to control, commercialise and monopolise everything. Western culture has a big mouth and very small ears. Even when it condescends to listen to other cultures, it does so within the framework of "searching for spices" and specifies both what it wants to hear and how it should be said. So it often ends up hearing only an echo of its own voice. It cannot stand unfamiliar talk or familiar but random talking. It worships order, format and precision. Since the Industrial Revolution, western culture has consistently laboured under what Robert C. Solomon has called a "transcendental pretence" (Solomon, 1980), that is, a strong impulse to present western ideas, whether optimistic or pessimistic, *sub specie universali*, if not *sub specie aeternitati*, and as the only rational and universally valid ones. As a consequence of this, perhaps, western philosophies and systems of thought have also been characteristically obsessed with such categories as "justification", "proof", "coherence", "consistency", "certainty", etc. But western intellectual history itself clearly shows that westerners have always been able to "justify", "prove", etc., ideas and practices which they later "unproved" and rejected when it was time to make use of new certainties, proofs, and justifications, etc. With enough motivation and sufficient effort, any proposition or point of view can be proved to the prover's satisfaction.

By contrast, African cultures, worldviews, systems of thought, religions and philosophies are characterised by diversity and, left to themselves, united in their tolerance and liberalism, live and let live attitude, non-aggressivity, non-proselytising character and in their accommodation of the most varied diversities and peaceful cohabitation of the most apparently contradictory elements. In many African households you would find, peacefully cohabiting under the same roof, advocates and practitioners of modern science, traditional medicine, doctrinaire Marxism, Capitalism, Catholicism, Protestantism, Bornagainism, Islam, etc.

So far, I have made a good number of what may be called generalisations and will be making even more in what follows. But I consider most of my generalisations to be of the type: "This is a bag of Bamenda beans" which does not depend on the absence of counter instances for its truth. There is a certain almost reflex-action practice within western scholarship, derived perhaps from formal Aristotelian logic , whereby the first impulsive reaction to a statement of the above type is to immediately reach inside the said bag and search for a pebble and then brandish it triumphantly as a "disproof" of the claim. A few pebbles do not "falsify" a bag of beans and, in this particular case, in fact, there could scarcely be a bag of genuine Bamenda beans without a few pebbles among the beans. Similarly, "Lying is morally wrong" or any other such moral judgement, is compatible with putative exceptions without loss or even diminution in its truth value and/or universality.

Although the examples I will be using in this paper are mostly drawn from my own natal background, among the Nso' of the Bamenda Highlands of Kamerun, I have no hesitation calling the perspective I am trying to present an African perspective. Something could not be correctly described as being Nso' without necessarily also being Kamerunian and, *a fortiori*, African, although the reverse may not necessarily be the case. Just as I am first and foremost an African before being a Kamerunian from the Nso' Kingdom of the Bamenda grasslands. It is, however, no secret that Africans, in spite of remarkable diversities, do have a common outlook on life, a common world-view and similar philosophics and practices comparable to the western world-view, outlook, philosophies and practices. Kwasi Wiredu (1980, chapter 1) has drawn together the underlying common elements and features of traditional African culture. No African, I believe, would disagree with him although it is possible to add new items to the ones he has highlighted and to suggest a qualification here and there. There are, of course, considerable diversities and differences amongst Africans. But we can talk of African culture in exactly the same sense that we talk of western culture. When we talk of western culture, we are justifiably bracketing, among many others, peoples in some ways as different as the French, with their boldly overt neo-colonial policies, the Germans, with their remarkably high degree of efficiency, conservatism and public conformism, the English, with their evasive equivocations and preference for **indirect** actions, and the Americans, with their highly loquacious, open and media-centred approach to all issues.

TRANSLATING AND INTERPRETING

In trying to present an African perspective of Bioethics, especially where this is nowhere to be found in written texts, one is necessarily assuming the role of a translator and interpreter, in the widest sense. In doing this, one must keep firmly at the back of one's mind Quine's famous Indeterminacy Thesis (IT) (Hallen and Sodipo,1986; Hallen,1995) whose effect should be to help keep one's limitations ever in mind but which should, however, not be exaggerated. It is difficult but certainly not impossible to do this job of cross-cultural translation and interpretation. Admittedly, beliefs, value-judgements and abstract ideas may be more difficult to translate than more concrete ideas and factual descriptive statements. But I believe that something like a **significance principle** of **translatability,** by analogy and contrast with the famous logical positivists' "verifiability principle of meaning", might be true here. By this putative principle, the significance, as distinguished from mere meaning, of a statement, sentence or proposition, would be directly related to its translate-ability into other languages, so that the relatively instranslatable parts are also, relatively, the less significant ones.

Every language captures certain aspects or elements of reality much better than others. So there will always be meaningful expressions in every language (culture) which are not fully or completely translatable into other languages (cultures). But these, on my thesis, would usually be expressions of less general significance and, thus, relatively of less importance. It is hardly possible to transfer corn or any other similar grain from one basket into another without losing some grains. If it were cocoyams, it would be easier to accomplish the task without any loss. But if it were corn flour or any other flour, it would be even more difficult as some would be lost even to the very air. But it cannot be argued from this that it is impossible to transfer any of these things from one basket into another.

Let me give a quick example of what I mean. Take the following statement: "Lunch is ready; please, come to table; enjoy your meal!" I believe that what is significant and important in this statement can be translated into all other living (and even dead) languages although in some languages, such as my own mother language, Lamnso', no word exists for "table" or for "lunch" and no equivalent expression for "enjoy your meal" which translates even better than the original into German as "*guten Appetit*" and, similarly, into French as "*bon appetit*".

What the translator-interpreter requires is as good a grasp as possible of **both** of the languages one of which is being translated into the other. Lack of such grasp of **either** of the languages could make the task appear near-impossible. In the above example, the Lamnso' translator-interpreter would be completely defeated if s/he didn't know that "lunch" simply means "afternoon meal" and that "come to table" does not mean that you should come to a table, even though the meal is laid on a table, but is simply an idiomatic way of inviting a person to participate in a meal. Further awareness that, within Nso' culture, no one needs an invitation to partake in a meal taking place in his/her presence and that, in all cultures, good cooking is only to be known in the eating, would help the Lamnso' translator-interpreter to render the above statement, *salva veritate*, into Lamnso' simply as: "Vifa veyi vi si tiyi" (Food is ready). Lack of idiomatic sensitivity can result in very misleading translations. And with African languages which are tonal, even the wrong inflection on a particle can change the meaning of a sentence completely. P.M. Kaberry, the pioneer ethnographer in Nso', was misled by the wrong stress on "*ve*" to translate "*viriim ve shuiy*" (incest) as "witchcraft of the sun", whereas the correct meaning is "daylight witchcraft" by analogy with "robbery in broad daylight". Excusably. Less excusably, the trio, Chem Langhee, Chilver and Fanso (1985), translate "*tee shishur she ngwerong*" as "stir ngwerong's pepper" whereas the idiom-sensitive translation is simply "prepare food for ngwerong".

NSO' METAPHYSICAL CONCEPTIONS AND WORLDVIEW

What I am attempting to present here is very much subject to my own limitations as a translator-interpreter and liable to great improvement by those more qualified than I. It may just sound like a ramble and the system described may not be entirely coherent. I have presented the ideas as they occur to me and, rather than try very hard to impose coherence on them, I have left them as they are and tried to draw some conclusions from them. As noted earlier, African plural systems in all spheres always accommodate some incompatibilities, incoherence and contradictions. May be that is the price any plural system of great diversity has to pay. However, the situation does underline uncertainty, human epistemological limitations and should foster patience and tolerance.

Nso' Bioethics can be said to be co-extensive with their Ethics and Ecology in such a way that it may be misleading to simply qualify their ethical concerns as being either eco-centred, bio-centred or anthropo-centred. Within this system, the centre would seem to be in constant motion in a manner similar to the fact that a straight line can be drawn from any chosen point on the earth's surface to its centre. What is indisputable, however, is that the three evolutionary levels: floral, faunal, human, are recognised, as well as the fact that the moral outlook and moral imperatives are limited to human beings, out of all the other entities with which they share the universe. The moral outlook and moral responsibilities, then, are anthropocentric but not morality itself. The Nso' certainly do not look on themselves as privileged creatures with a God-given mandate to subdue, dominate and exploit the earth and the rest of creation. They do, however, consider human well-being as the aim, end and purpose of morality.

The practice of **shifting cultivation** which, unfortunately is fast disappearing, owing to both increasing population and increasing use of imported chemical fertilisers, was one way of ensuring that the soil, vegetation and the fauna recovered and reasserted themselves after a period of human exploitation. Strict taboos on eating or killing certain insects, reptiles, birds and mammals also ensured the survival of the fauna within this ecological niche. Taboos are a particularly effective deterrent because the consequences of their violation, whether deliberate or involuntary, are metaphysical, affecting not only the life of the violator but his/her very being as well as the entire community.

For the Nso', non-human animals in general can be divided into three groups: (1) Wild animals (*nyamse kwa'*), the non-tabooed species which can be freely hunted and eaten subject only to surrendering certain specific ones such as the leopard (*baa*), buffalo (*nyaar*) to the King or to *Ngwerong* in the case of the spiky porcupine (*kigum ke boo*) or the boar-constrictor (*ngvem*) - (*nyamse ngwerong*). Killing a pregnant animal is, however, strictly taboo and requires immediate ritual cleansing and purification, even if done accidentally. (2) Livestock reared purposely and specifically for food and other human needs, such as fowls, goats, sheep, rabbits, cattle etc. Of these, fowls and goats (particularly the castrated he-goat) are the most indispensable because they are used in myriad ceremonies and celebrations (marriages, births, deaths, installations, rituals, initiations, etc.) and as common gifts. It would not be an exaggeration to say that, if fowls and goats, taking along with them maize, red palm oil, salt and kolanuts, were to go out of existence, Nso' culture, as we know it, would disappear with them. (3) Domestic animals kept or reared as human **helpers** (not **companions** - for companionship the Nso' generally rely only on other human beings) such as dogs, cats, horses, donkeys, etc. It is strictly taboo to eat any of these. They are treated with care, sensitivity and compassion but you are unlikely ever to see a Nso' person lovingly cuddling a dog or cat or kissing a horse as is not uncommon in western cultures.

Among human beings themselves, relationships are underlined by the absolute equality of all human beings *qua* human beings and consequent radical respect of individuality as seen, for instance, in the fact that each child is believed to come into the world with its own name which the parents or family only try to guess (*shu' yir wan*) and the fact that, before western colonisation, a married woman did not bear her husband's name nor a child the father's name, although both father and mother could be renamed after the child (Mzeka, 1993; Banadzem, 1986). The weak, the poor, the deformed, and the physically handicapped are particularly respected because of the belief that such persons usually possess extraordinary "depth", have unusually powerful personal spirit-gods, or are frequently used as disguises by the spirits or by God.

For the Nso', the earth (*nsaiy*) is a very potent force, to the extent that they do not cut the earth lightly (without ritual permission). In Nso', the hoeing (as well as the planting and harvesting) season is always ritually inaugurated after appropriate sacrifices and incantations, performed by the King (Fon), the chief High Priest of the Kingdom, assisted by his three ritual helpers: '*Tawong*' (father of the Kingdom), '*Yewong*' (mother of the Kingdom) and 'fai_Ndzendzev' (chief great councillor of the Kingdom) (Kaberry, 1952; Mzeka, 1978). The earth is where living things - plants, animals and human beings - originated and where they all end up again. (Thou art earth and into earth shall thou return). In terms of awe and reverence, the Nso' stand in awe and reverence of the earth, more than the entire plant kingdom, the animal kingdom and the world of human beings. The law of the earth (*nser nsaiy*) is the strongest and brooks no breaking because the consequences of its violation are severe, metaphysical and unavoidable. The earth is the final arbiter of all human affairs. The Nso' year (*ya' Nso*) always begins with fertility rites (for plants, animals and humans) which are addressed to the earth and which terminate with the ritual cutting of the earth (*gbar nsaiy*) with the sacred hoe of the Kingdom (*kisoo ke wong*).

As E.M. Chilver (1990, pp.239-240) has remarked concerning Nso' conceptions about the earth, "...the earth was sometimes spoken of as a separate being with a will and body (*wun*) of its own, able to afflict perjurers in disputes over land and property, inflict bad death on secret evil-doers and even to pass judgement on kings". And as Kaberry (1952, p.33) had observed much earlier: "Human quarrels and transgressions affect fertility in its widest signification....The tie between '*nyooiy*', the earth, and the people who live on and cultivate the earth, is a close one and is expressed in moral and ritual terms." Divinities or god-spirits (both benevolent and malevolent) and ancestor spirits could reside in a rock, waterfall, lake or tree and can, from time to time, assume human or animal form. The departed ancestors reside in the earth as well as evil spirits (*vibai ve bivi*) and benevolent spirits (*vibai ve jungvi*). It is less clear where exactly God himself resides but there is no concept of a Christian-type heaven. However, libations to God are poured on the earth and prayers to him/her are addressed to the earth as a whole. The dead depend on the actions, especially ritual sacrifices, of the living for their well-being and the living, in turn, depend on the solicitations and intermediacy of the dead for their health, progress and well-being. It is very significant to note that the conflict between the two leading family lineages in Nso' (Mzeka, 1990:p.115f; Kaberry's Fieldnotes) which started in 1925 and has continued intermittently over several "settlements" and "reconciliations", originated from the refusal of the King (Fon Ngga Bifon I,1910-1947) to perform traditional ritual sacrifices following the death of Fai Ndzendzev, Toombu (died 25 January 1925), and consequent illness in the Ndzendzev family.

Supernormal human beings (*wir sem*) can also transform themselves into other creatures. '*Sem*' (or 'depth' literally) is, however, generally, of two types: benevolent (kingly) "depth" (*sem vifone*) and malevolent "depth" usually associated with witchcraft and sorcery (*viriim*). (Banadzem, 1986:pp.182f; Chilver, 1990:pp.232f). The King (Fon) of Nso' is particularly believed to be capable of transformations as an aspect of his kingly attributes. During the so-called "punitive expedition" when the Germans invaded Nso' in 1906, and the ensuing Nso'-German war (April 27-June 5,1906), the inability of the Germans to capture the Fon (Seem II, 1880-1907) until he himself voluntarily accepted to surrender, was widely attributed to his transformative powers (Mzeka, 1990: p.78f; Fanso and Chilver, 1995: p.17). One of the most charismatic Kings (Fon) of Nso', (Seem III, alias Mbinkar Mbinglo, 1947-1972), in one of his boastful moments, declared that he had 27 possible different transformations (*vibai*).

Within a western framework, the most natural reaction here would have been, perhaps, to immediately go for pen, paper and camera and try to persuade him with monetary offers to prove his claim "scientifically". But in this context, that would really be like asking if you could be allowed to attend the Holy Mass of African Traditional Religion to see things for yourself and if you could, please, be told the order of service and liturgical readings in advance to enable you prepare for the Mass. That is simply not the way things work within this system. Within this system, super-human and even extra-ordinary knowledge and capacities/capabilities are looked upon with awe and even dread because of the heavy moral responsibilities they impose and the dangers of abuse. When at birth a child was known or suspected of possessing such endowments, (*won nyuy* - spirit children), special rituals were always performed to ensure that they would be used benevolently and not malevolently or else to exorcise the child of them. The same thing happened in the case of children who, though delivered as normal children, are later "stolen" by a spirit (*nyuy shong wan*) and go missing for a few days only to reappear with very unmistakable signs of what had happened to them (Mzeka, 1993: p.20f; Banadzem, 1986: p.182f).

Furthermore, ordinary human beings, if they are very ambitious or misled by others, could seek these extraordinary psychic powers through a process called "*du kwa' sem*". They could then acquire the power, *inter alia*, of transformation into various other creatures - flies, insects, birds, reptiles and mammals. And sometimes they try to lure innocents into their enchanting world. So, you have to be very careful in the way you treat all living creatures, human and non-human, including the lowliest, because your own child or parent

~ 18 ~

could transform into a non-human creature and even God himself could appear to you in the guise of a stranger without your knowledge. The important thing here, of course, is not so much **veridicality** as the very fact that these beliefs and claims are part and parcel of a complex **world-view and belief system**. E.M. Chilver, who joined P.M. Kaberry in her subsequent researches in Nso' and continued after the former's untimely death in 1977, informs me that, while discussing Fons' names with Seem III alias Mbinkar Mbinglo (1947-1972), he had told them that the Fon can change himself into a lion (*bvere*), but had quickly added that that was only "*fihsin*" (i.e. a metaphor). The quick caveat might or might not have been calculated to pre-empt a request for a demonstration. In any case, it is taboo for Nso' people to kill a lion. When Nso' people, especially hunters, encounter a lion, they respectfully clap their hands the way they usually clap to greet the Fon, and the lion walks away majestically.

Thus, the Nso' world-view is, first and foremost, communitarian. But it goes beyond anthropological communality, even goes beyond biocommunitarianism (Callicott, 1994 p.156f.) and may be more appropriately described as eco-bio-communitarianism. Nevertheless, Nso' morality is ultimately and fundamentally human-centred in so far as its teleological end and limits are defined by human well-being. Super human spirits and non-human animals are not prone to moral error nor are they vulnerable to the consequences of their own moral errors. In this way they are different from human beings in not being subject to moral imperatives and sanctions. They can, unlike human beings, afford to act arbitrarily. The earth (*nsaiy*) holds no threat for them.

NSO' POSITION ON SOME BIOETHICAL ISSUES

As already stated, Nso' Ecobioethical conceptions may not be entirely coherent, neat and tidy, but they do give a rough plastic framework within which particular actions can be evaluated and justified or condemned. It is doubtful that real life, anywhere, can ever or even ought to be as neat and tidy, as coherent and consistent as theory-making would want us to make it. The fact that the Nso' greatly relish "*kiban wuna nyu sejisi*" (foofoo corn and njamanjama) as a staple, for instance, does not mean that they do not sometimes or even frequently eat roasted cocoyams. Real life is more like a Bamenda gown (*vikum ve Kom*). If you "gather" only one side of it, it is unbalanced, giving you the funny look and inequilibrium of a courting cock dancing on one leg with one wing lowered to the ground. You also need to "gather" the other side and then it is well-balanced, allowing you to step forward with a firm stride, grace and dignity, like a paramount chief.

Conceptualisation and theory-making are like wall-building. If the wall is built with concrete, it protects the house or city very well and keeps out intruders most effectively. But, if one untypical day the house-owner or *bona fide* city-dweller himself/herself has cause to want to get out quickly, in an emergency, for instance, the very wall, which hitherto served as protection may become the greatest danger and obstacle to a safe escape. And that is when the advantages of plasticity in concept and theory construction may become evident.

From this point of view, one problem with western ethical thinking (and western thinking in general) is its obsession with details and the incorrigible conviction that there is one correct principle from which everything is deducible and to which everything is relateable, as well as the perennial search for such a principle. This conviction is compatible with and, in fact, leads naturally to undue exaggeration, dogmatism, extremism and intolerance which, in turn, engender crusading and fanaticism. Nso' views on some of the bioethical issues that are most hotly discussed in the western world, would be characteristically devoid of minute details as well as extremism and dogmatism though, nevertheless, quite firm in their general outline. Here I can attempt no more than a brief and rough indication of a Nso' approach to a small randomly selected number of these issues. I am attempting only a broad overall picture or framework within which the sort of detailed analyses and minutiae which interest most western bioethicists can be filled in on a case by case basis as required.

It should, however, be stated that **some** of the bioethical issues currently most discussed in western society, such as those which arise directly from very recent technological developments, would not make much sense to an average Nso' person, although that is not to say that a firm opinion on them, in line with overall Nso' metaphysical and ethical conceptions would be difficult to formulate.

Generally, it can safely and truly be said that the Nso' value children so highly as to consider procreation one of the main purposes of life. Every Nso' person would prefer his/her own death to that of his/her child. '*Bongkpumo*' (I prefer being the one to die) is

~ **20** ~

a very common name in Nso'. The Nso' have never known how to take the death of a child or young person. Distress at such an eventuality always seems inconsolable. S/he is usually quickly buried without ceremony but along with every personal effect that might help to remind people of him/her and then quickly forgotten. By contrast, an old person, especially one with many children, is always buried with elaborate ceremony, celebration and even rejoicing. The Nso' also value well-being so much that they prefer death to irremediable suffering. '*A si ngeh bong kpu*' (Death is preferable to suffering) is a common Lamnso' saying. The early Christian missionaries in Nso' really had a formidable doctrinal weapon for conversion in their concept of **hell**: a place where those who refused conversion would suffer deathlessly, *per omnia secula seculorum*! And, in spite of the very abundant harvest of converts and subsequent very healthy growth of Christianity in Nso', it can be said that the doctrine of **hell fire** has remained one of the most incomprehensible dogmas of Christianity to an average Nso', as can be very clearly seen from Kenjo Jumbam's only slightly fictionalised autobiographical novel *The White Man of God* (Heinemann, African Writers Series, No. 231). The Nso' also value the community so much that they subordinate the individual to the community. They also constantly emphasise their own limitations, ignorance, moral weaknesses and culpability vis-à-vis other human groups. Almost every second Nso' person is likely to be bearing one of the following names: *Nso'kika* (what do Nso' people know?), *Verbe* (where are we?), Verjai (we erred), or *Yiinso'batoiybarah* (we Nso' people are at fault though we are blaming modernity (others) etc.

EUTHANASIA, SUICIDE AND ABORTION

The Nso' would probably approve of **euthanasia**, understood in its etymological meaning as **a gentle and painless death**, to the same degree that they would disapprove of **suicide** or **abortion**, understood as the **termination of a pregnancy by deliberate human intervention before the embryo or foetus has reached term or viability.**

For the Nso', a good death would be defined as a relatively painless one that is neither pre-mature nor overdue. A death is not considered pre-mature if the dying person is old enough to have begotten adult children, provided wives for all his male offspring (which is one of his duties as a parent), witnessed the marriage of all his daughters (for some of whom it is **not** his responsibility to provide husbands)[3], married off all the first daughters of all his own daughters (which is also one of his duties), seen at least some of his own grandchildren, and kept the family and lineage united, healthy and happy. A Nso' person with such a profile looks forward to death with great composure and sometimes even a bit of impatience. These achievements are not, of course, entirely dependent on chronological age and, for the Nso', chronological age is less important in being considered an elder (*nggaywir*) than achievements. Your younger brother effectively becomes your "father" if he is selected and installed the lineage head[4]. Similarly, your sister or even daughter could in some circumstances become your "mother".

Whenever a Nso' elder who has accomplished his/her mission in life falls sick, s/he would always pray that, if his/her time has come, God should take him/her speedily. And if s/he dies, mourning is mainly a ritual ceremony tinged with rejoicing and accompanied by much dancing followed by feasting. In 1960, Kaberry and Chilver attended the burial of one of the daughters of the late Fai Shindzev - a second degree prince (*wanwan-nto*) in Nso' - and were greatly surprised to discover that, while some women were mourning and wailing, the mood of the grave-diggers nearby was a rather joyous and hilarious one. Upon enquiry, they were told: "They dance with joy because the woman who has died is one who is an elder (*nggaywir*), who has many children" (Kaberry and Chilver, *Nso' Working Notes*, 1960, pp.1&2). At such a ripe old age, the Nso' fear illness and suffering but not death. It is thus clear that any Nso' person of that description would seriously object to being kept on such a thing as "life support" in case of terminal illness.

---

[3] According to Nso' custom, the first daughter of every properly married woman is given out in marriage by the daughter's maternal grand-father (*kitaryiy*).

[4] Once in 1988, I accompanied a friend and his youngest brother, who was living with him in the city as a dependant, to the burial ceremonies of their father in Nso'.On getting there, they seized our young companion, who all along had been acting as a sort of servant to both of us, and installed him as their late father's successor and head of the family lineage (*nyilah*). Our relationship to him dramatically changed. At the end of the ceremonies, we sought audience with him to ask for permission to return to the city. He compelled us to spend an extra day before permitting us to depart.

Voluntary euthanasia, in these circumstances, would therefore be quite unproblematic for the Nso'. But, since they fear suffering more than death, it is also clear that the Nso' would welcome **non-voluntary euthanasia** for all categories of certified terminal patients, where all that can be expected is pain and anguish for both patient and loved/loving ones but where the patient, for any reason, is incapable of making a choice. It is, however, quite difficult to see how any case of **involuntary euthanasia**, where the patient, though capable of doing so, refuses consent, could arise within this context. If a terminally ill patient refuses euthanasia then, clearly, either his/her suffering is not insupportable or s/he prefers insupportable suffering to death. In either case, there would be nothing further that can be done than respecting the patient's wishes. Simply saying nothing, in this case, would be exactly equivalent to expressly refusing because, within Nso' culture, in the face of a question or suggestion, silence does not signify consent as it might within other cultures, but rather refusal or disapproval. But, generally, the Nso' find no reason for clinging on to life artificially.

All this accords rather remarkably well with one of the conclusions drawn by Burgess and Tawia (1996, p.24) from their recent study on the beginning of human consciousness, to wit: "...the view that pain and suffering are intrinsically bad probably comes as close as any to universal acceptance in both unreconstructed commonsense morality and in systematic normative morality." The Nso', in any case, certainly consider pain and suffering as being intrinsically bad.

By contrast with euthanasia, **suicide** is strictly taboo to the Nso' and is considered an offence against the earth (*nsaiy*) requiring very elaborate ritual purification. Some western bioethicists have managed to forge a link between euthanasia and killing and suicide in the ideas of **"assisted suicide"** and **"mercy killing"**. These coalesced ideas cannot fit within a Nso' bioethical framework. Within that framework, these ideas are quite contradictory because euthanasia is euthanasia precisely because it is neither suicide nor murder. For the Nso', both suicide and murder are abominable and to commit or assist commission of any of them is, therefore, equally abominable. In 1963, Fai Taakiwuf, then one of the oldest people in Nso', described suicide to P.M. Kaberry as "*kinyu ke ko'oiyinki*" (a horrible abomination) (Kaberry, *Manuscripts and Fieldnotes*, 29.V111.63). His reasons were that a person who commits suicide shows contempt and disregard for the family and the community by not giving "*ntir kpu*" (death bed advice) and blessing them before dying, as should be the case. S/he dies "whole" (*wu kpu i fur*) without indicating a successor, without indicating his/her unpaid debts, unsettled quarrels and disputes, unfulfilled promises etc,- all matters of great importance on which the health, well-being and prosperity of the family and the community depend. He can therefore be truly said to want to destroy those he is leaving behind.

So while it is morally acceptable for a terminally ill patient to seek and to be helped to find a gentle and painless release from meaningless pain and suffering, it is morally unacceptable, otherwise, for a human being to terminate his/her own or another's life.

Regarding **abortion**, the Nso' attitude can be said to be generally guided by the taboo against harvesting premature crops or fruits. In Nso' it is taboo for people to harvest any crop before it has been certified mature enough for harvesting and appropriate ritual ceremonies performed. Clandestine plucking of immature crops or fruits is believed to bring personal misfortune. It is for the same reason that sex with minors (before adolescence) is taboo. Sometimes a girl was betrothed to a husband as early as when she was still 5 years old. As she grew up, she would spend lengthy periods in the husband's compound or sometimes, in fact, she moved permanently to live in his compound. It was, however, taboo for him to have sexual relations with her before maturity.

I have already mentioned above the taboo against killing pregnant animals. It is obvious that it is the unborn baby which makes the taboo necessary. It is the same with human beings. In a recent letter to me (dated 21/01/1996), E.M. Chilver told me:

> *I recall being thunderstruck by a conversation with Fon Sem III Mbinglo about the execution of persons guilty of treasonable adultery, a capital offence everywhere. The 'wiynto' (Fon's wife) was guarded by her lineage-head if pregnant, lest she hanged herself, and then clubbed as soon as she had given birth. Even though I knew, by that time, what sort of reply I would get I asked: "Why not kill her sooner?" The Fon, deeply shocked by my barbarity, replied: "What Nso' man would pluck an unripe melon!?"*

The Nso' position is thus clearly against deliberate non-therapeutic abortion. Therapeutic abortion would, however, have been justified.

If the life of the pregnant woman was at risk because of the pregnancy, it would have been considered that an evil spirit had, perhaps, disguised itself and lodged in her womb under pretext of being a child. It would have been dislodged through appropriate rites and medicaments.

The Nso' position on abortion, however, has some paradoxical consequences. It has the consequence that, while non-therapeutic abortion was considered wrong, **infanticide** was sometimes considered permissible. This happened in the case of **incest**. In Nso' incest, *'viriim ve shuiy'*, (broad daylight witchery) is strictly taboo. The Nso' are one of the very few (perhaps the only) African peoples who pay neither bride price nor dowry for marriages. Nevertheless, marriage is a very elaborate affair. And most of the concern is to make sure that there are no unacceptable consanguinal links between the potential spouses. Unlike amongst many other peoples, the Nso' do not tolerate marriage between cousins, up to the fifth degree. In fact, incest, in Nso', is mainly understood as sex between close cousins. Biological-parent/child incest and incest between direct siblings is almost unthinkable in Nso'.

Incest, which is believed to pollute the entire family and lineage, requires a really frightful purification ritual. A really good example is narrated in Kenjo Jumbam's *The White Man of God* (chapter 11). The two offenders are led in a solemn ceremony to a public road junction and made to simulate coitus on a new mat laid on the ground, while their abomination is ritually symbolically transferred unto a he-goat which is then clubbed to death, wrapped up in the mat and buried in a very deep grave. Thereafter, the stigma is completely removed and the woman develops the pregnancy quite normally. But, when she delivers, the child would be immediately killed and the news would go round that "as she was putting down her pot from the fire it fell down and spilled its contents but that she herself was not hurt." The child is then quickly buried and it is taboo for anybody to weep or even as much as sniff the air.

CONCLUSION

It is clearly up to western Bioethics and western systems of thought and practice in general to allow African Bioethics and African culture in general to influence them. If only more westerners could really honestly try to get into the spirit and swing of things African[5], in the same spirit that many Africans have honestly and enthusiastically got into the spirit and swing of things western, humankind and the entire biological world might stand to reap great benefits. Africans have tried. From western Christianity through western languages and education, to western systems of thought, philosophies and fashions, Africans have honestly and enthusiastically got into the spirit and swing of things western. In the process, Africans have benefited from western culture and used it to enrich their indigenous cultures. But, unfortunately, in so doing, Africans have also neglected some vital aspects of their own indigenous cultures which could, in turn, have helped to humanise and enrich western culture. As there is no possibility of Africans imposing these putative benefits of African culture on westerners through any putative "black man's burden" and "decivilising mission", it is really up to westerners to salvage these elements of African cultures for the enrichment of western culture and the benefit of humankind, since western culture is, indisputably, the overwhelmingly dominant culture of our historical epoch. Lastly, I should say a word about my constantly shifting tenses. I am quite aware that the tenses of my narrative have been shifting back and forth between the present and the past. This clearly reflects the African predicament and the struggle **for** and **of** the African soul which began with European colonisation and has continued up to the present. The relative insularity, consistency and homogeneity of the

---

[5] A few westerners who have made an honest effort of this kind are to be found among some of the missionaries. The only secular westerner known to me personally who took such a plunge is Ulli Beier. In recent conversations (to be published) with Wole Ogundele, Beier testifies regarding his contact with an African culture: "...one cannot say that one culture is greater than another ... but ...there are certain cultures which have a special kind of luminosity, and perhaps Yoruba culture is one of them ... that is what I always felt. To me it just shone, radiant and expansive, no pomposity, no self-consciousness. I got a whole life from it. I did not go into it thinking there was something I could add ... that would have been ridiculous. My quest too was not that of a researcher, with a tape recorder or pencil and paper. Indeed, it would have been sacrilegious ... That was impossible for me ... Life is to be lived, not explained away. I never could bring that bad aspect of European culture into my association with Yoruba people and culture."

Nso' world-view, customs and practices was first jolted and violated by the arrival of the German colonisers in 1902. This was followed by the imposition or introduction of European languages, systems of education, law and, above all, Christianity. Ever since, there has been a continuing and undecidable tug-of-war between orthodox Nso' culture and way of life and an emerging syncretic culture and way of life. In this situation, it becomes necessary to speak of traditional practices which have been abandoned, are in abeyance, or which are becoming rarer and rarer, in the past tense, even though the chances are there that they could still reassert themselves.

REFERENCES
Published Works
Burgess, J.A. and Tawia, S.A. (1996): 'When did you first begin to feel it? -Locating the beginning of human consciousness' in *Bioethics*, Vol. 10, No. 1, January 1996.
Callicott, B. J. (1994). *Earth's Insights: A Survey of Ecological Ethics from the Mediterranean Basin to the Australian Outback.* Berkeley, Los Angeles and London, University of California Press.
Chem-Langhee, B. (1985): 'Nto' Nso' and its Occupants: Privileged Access and Internal Organisation in the Old and New Palaces', in *PAIDEUMA*, 31(1985). Page number?
Chilver, E. M. (1990): 'Thaumaturgy in Contemporary Traditional Religion: The Case of Nso' in Mid-Century', in *Journal of Religion in Africa*, xx,3. Page
Hallen, B. (1995): 'Indeterminacy Ethnophilosophy, Linguistic Philosophy, African Philosophy, in *Philosophy*, 70/1995, Page
Hallen and Sodipo, (1986): *Knowledge, Belief and Witchcraft: Analytic Experiments in African Philosophy.* London: Ethnographica Publishers.
Hountondji, P. (1983): *African Philosophy: Myth and Reality.* London, Hutchinson.
Jumbam, K. (1980): *The White Man of God.* London, Heinemann.
Kaberry, P. M. (1952): *Women of the Grassfields: A Study of the Economic Position of Women in Bamenda, British Cameroons.* London: Her Majesty's Stationary Office.
Mzeka, P. (1980): *The Core Culture of Nso'.* Agawan, Ma., Jerome Radin Co.

------------ (1990): *Four Fons of Nso', Nineteenth and Early Twentieth Century Kingship in the Western Grassfields of Cameroon*. Bamenda: The Spider Publishing Enterprise.

----------- (1993): 'Rituals of Initiation of the Western Grassfields of Cameroon: The Nso' Case", in Mbunwe-Samba, P. et. al, *Rites of Passage and Incorporation in the Western Grassfields of Cameroon*, Vol. 1, Bamenda: Kaberry Research Centre.

Oruka, O. H. (ed.), (1990): *Sage Philosophy: Indigenous Thinkers and Modern Debate on African Philosophy*. Leiden, E.J. Brill.

Singer, P. (1979): 'Unsanctifying Human Life', in Ladd, J. (ed.): *Ethical Issues Relating to Life and Death*. Oxford: Oxford University Press.

-------------- (1993): *Practical Ethics*. Cambridge: Cambridge University Press.

Wiredu, K. (1980): *Philosophy and an African Culture*. Cambridge: Cambridge University Press.

Wright, R. (ed.), (1979): *African Philosophy: An Introduction*. Washington: Washington, D.C., University Press.

Unpublished Sources

Banadzem, J. L. (1986): 'Nso' Kingdom; From the Installation in Kimbo to Ngga Bifon 1. Political System and Ideology (1780-1947)'. Thèse de doctorat de 3e cycles. Université de Paris 1.

Fanso V.G. and Chilver E.M. (1995): 'Nso' and the Germans: the First Encounters in Contemporary Documents and in Oral Tradition'. Draft Mimeograph.

Kaberry, Phyllis, M., Manuscripts and Field notes, Archives Section, London School of Economics (LSE).

Kaberry, P.M. and Chilver, E.M. (1960): 'NSO' WORKING NOTES: Some items selected from Dr. Kaberry's and E.M. Chilver's notebooks of 1960 by E.M. Chilver, Mimeograph.

CHAPTER THREE

THE ABORTION DEBATE: ETHICS, CUSTOM AND LAW IN INTERACTION

[This chapter was first published in the Journal "Biodiagnostics and Therapy, Magazine Bilingue de Sante Publique au Cameroun" in March 2002]

PREAMBLE

The last quarter of the 20th century witnessed a significant shift of emphasis in global philosophical trends from concern with overly theoretical issues to more practical matters. Bioethics, the study of the ethical, social and legal issues arising from existence, life and the biological sciences, is one of the fruits, among many others, of this shift in philosophical emphasis. It has brought philosophers down from the ever shifting clouds of sterile speculation to the firm ground of life and death issues and, at the same time, raised ordinary men and women of the mundane world to the realms of inevitable philosophical reasoning, if not speculation.

That human beings are at the apex of biological existence is a claim that is sometimes dismissed as too arrogant a claim and has often been contested by some militants of the rights of plants and/or animals. But this claim can be considered merely as a perceptible datum, from the point of view of morality. One way to easily realize this is to consider the fact that, while human beings have putative moral responsibilities towards plants and animals, these latter cannot be considered, without absurdity, as having any reciprocal moral obligations towards humans. The consequence of humankind´s position as the summit of biological life, as we know it, is that it bears alone on its shoulders the whole weight of moral obligation and responsibility for the biological world.

Reproduction is crucially important for all living things. All living things reproduce themselves and also die. Without reproducing, living things would go out of existence. Reproduction is thus a central area of bioethical concern and human reproduction can further be considered as the nucleus or epicentre of that domain, since human beings bear the whole weight of moral responsibility for the entire biological world. That responsibility may justify human intervention in the biological world, although that does not imply that every conceivable intervention is justifiable.

In this chapter, I consider the main positions for and against abortion. I argue that ethics, laws, customs and religious injunctions need to be constantly revised in the light of reason. Further, I argue that, if we start from ordinary, disinterested, commonsensical reason, the rationally most defensible position on abortion lies somewhere midway between the extremes.

INTRODUCTION

Of all controversial issues within human reproduction, there is scarcely any which generates so much heat, fervour and interest as **abortion**. This, indeed, is an issue on which many people seem to have a vested interest, to take a strong stand, rationally justifiable or not; an issue over which the emotions, non-rational instincts and extra-ethical considerations may easily put reason to flight. It is also here, more than anywhere else, that people may be prone to **akrasia** or moral backsliding, that perplexing condition in which a moral agent knowingly and willingly does what s/he knows and accepts s/he ought not to do.

The abortion debate has been greatly complicated by the existence of divergent ethical convictions, cultural practices, customs, laws, religious dogmas and economic motives. The **United Nations International Conference on Population and Development** (Cairo, Egypt, 5-13 Sept. 1994) greatly helped in bringing out some of these divergences and contradictions into play, before, during and after its work.

An ethical conviction is not necessarily equivalent to what is morally right, although it is justifiable for the individual always to act according to her/his ethical convictions. That may sound contradictory and has the consequence that one may act according to one´s ethical convictions and still be morally wrong in one´s action. Let me try to clarify the issue by making the following terminological stipulations. Ethics is a subset of **morality** that can be considered as being concerned most generally with right and wrong in relation to human conduct. Morality as such is based on human **reason or rationality** (the ultimate means by which right and wrong may be known) and the social nature of human beings. Freedom and the ability to make choices are essential for morality. Rationality and sociality are the defining and distinguishing characteristics of the human person and freedom defines the sphere of morally responsible actions.

Human beings are, by nature, both rational and social. However, rationality and sociality are **necessary** but not **sufficient** conditions for morality. Freedom completes the conditions which, taken together, are sufficient for morality. Any creature which does not enjoy the use of reason or which is not a social being, an egregarious being, or which is not free to choose between alternative courses of action cannot be subject to morality. Freedom is the point of departure for morality in the sense that without its assumption, rationality and sociality notwithstanding, the notions of **right** and **wrong** in human conduct would clearly be absurd. Freedom of action is the irreducible basic assumption for morality, for responsibility and culpability, for the ascription of praise/blame, the apportionment of reward/punishment, etc. To predicate any of these things of an agent is to assume that the agent in question could have done otherwise in the given circumstances; that is to say, that s/he acted freely. So it is rationality, sociality and freedom, taken together, that give us the moral perspective or dimension and that impose limitations, obligations and moral imperatives on us. Now, since rationality, sociality and freedom are defining characteristics of human persons, it goes without saying that morality is or ought to be fundamentally the same for all human persons, everywhere, at all times.

The same does not apply to ethics. **Ethics** derives from morality and refers to a set of general principles by reference to which any relevant concrete human actions can be judged right or wrong and, therefore, moral or immoral. Moral discourse, articulate moral reasoning and argumentation belong in the sphere or ethics. While morality may be considered as being the same for all human beings, irrespective of time and space, ethics is, no doubt, partly a function of culture. Ethics is culture-tinted and may be affected by both time and place. A time there was, for example, when "an eye for an eye and a tooth for a tooth" was a generally accepted ethical principle, within some cultures. But it was not at any time a morally right principle as, indeed, it came to be realized within the cultures in question. "Love your neighbour as yourself" which replaced the above *lex talionis* is an ethical principle that is likely to be reformulated in our time as it becomes increasingly clear that morality equally demands love for our non-neighbours, including plants and animals.

Morality can be conceived *sub_specie aeternitati*. But ethical principles should be constantly reviewed and debated and reformulated to rid them of the discernible tintings of time and place, of culture and of human folies and fallibility. It is clear that this task can never be completed, inasmuch as every human situation has its own egocentric predicaments.

Joseph Omoregbe (1979, p.3) has illuminatingly compared the relationship between Morality and Ethics with that between thinking and logic or that between religion and theology. Without thinking there can be no logic, just as without religion there can be no theology. Logic is based on thought in the same way that theology is based on religion. Thinking is the same for all human beings but it may produce slightly different systems of logic, just as one and the same religion may engender different theologies. Any system of logic must be measured against logical thinking as such, just as any theology must be measured against the religion. In like manner, any Ethics or set of ethical principles must be measured against commonsensical Morality and simple Rationality. In ethical discourse and debates, the attempt to "ethicalize" practice rather than conform practice to the moral is not one that is infrequently encountered.

Ethics also has a further derivative in *Ethos*, which signifies "a professional culture of morally-guided persons" (Häring 1972, p.16), a written or unwritten code of conduct relevant to a particular profession or occupation. If Ethics is culture-tinted, Ethos necessarily reflects the limitations of particular groups of individuals or particular professions whose main business is not concerned with Ethics or Morality. Every professional group has its own ethos, whether written or unwritten, which may be found wanting in part or whole when measured against ethical principles, morality or rationality.

It is unnecessary, I believe, to further labour to show that cultural practices, customs, positive laws, and religious injunctions can fall short of ethics and thus, *a fortiori*, of morality. These, no matter how widespread, old, venerable or revered, can be mistaken and must constantly be reviewed in the light of unencumbered reason.

WHAT IS ABORTION?

Abortion, generally, has been understood as the termination of a pregnancy, either spontaneously or by deliberate intervention, before the human embryo or foetus reaches "term" or viability. Beyond that, we should more appropriately talk of infanticide. Spontaneous or accidental abortion need not concern us here since it falls outside the realm of morality in that it is devoid of intentionality, purposiveness, and voluntariness; in short, it is devoid of freedom which is a necessary condition for moral agency. In other words, spontaneous abortion is something that *happens* rather than something that is *done* or made to happen. Morality is not concerned with happenings but only with purposive actions. Things which simply happen, fall within the realm of the "is" where causal explanation may be called-for as distinguished from the region of the "ought" where justification is called-for. As such, spontaneous abortion might be of interest to the purely scientific researcher but it cannot, in itself, be of much interest to the moral enquirer.

Regarding deliberate abortion, it is a matter of dispute as to when a foetus should be deemed to have reached term. Some people fix the age of viability at 28 weeks from the time of conception. Such, for instance, is the legal age of viability in the United Kingdom. Other people argue for a lower age of viability, some going as far down as 16 weeks. What is beyond dispute is that improvements in medical facilities and techniques would seem capable of continuously and indefinitely lowering the age at which the foetus can survive outside the womb. But this does not much affect the main arguments for or against abortion. The age of foetal viability may be of interest for the pure scientist but it has nothing to do with the moral worth or otherwise of the foetus.

DELIBERATE ABORTION

Deliberate abortion may be therapeutic or non-therapeutic. Therapeutic abortion is the deliberate termination of a pregnancy in the interest of the pregnant woman's life or health. Many people would have no hesitation recommending therapeutic abortion if the maternal life or health is seriously at risk. Other people, however, who take the view that all types of deliberate abortion are co-terminus with murder (foeticide), disapprove of even therapeutic abortion. Such people might argue as follows: "Murder is always wrong and should never be deliberately committed under any circumstances. It is therefore better to **allow the mother to die** than to directly murder another human being (the unborn child) to save her life". Such a view is implied in the Catholic Churche´s position on abortion. Nevertheless, I believe that therapeutic abortion is rationally easily defensible, provided clear and acceptable criteria are specified as to when the pregnant woman´s life or health can be deemed to be at great risk. And here, lay people cannot but depend on the specialized knowledge and verdict of medical professionals. Medical professionals are, of course, not infallible and are prone to errors. But, as long as they are thorough and honest in their investigations and pronouncements, fallibility is not an important objection. To err is human. All human knowledge is fallible and there is absolutely nothing human beings can do to render human knowledge infallible.

Non-therapeutic abortion is highly controversial. Three clear positions have usually been taken on non-therapeutic abortion. The **first** of these is the extreme view that the foetus is merely a part of the pregnant woman's body, over which she can exercise her "autonomy". Here there is talk about "every woman's right to have an abortion". The **second** position, which is the polar-opposite of the first, holds that the foetus possesses, in itself, an absolute and inalienable right to life which ought not to be violated under any circumstances. Such a position has consistently been defended by the Catholic Church through the ages. The **third** position, which threads trippingly between the above two extremes, is the view that non-therapeutic abortion may, under certain circumstances, be justifiable.

THE EXTREMES

There are people who hold the view that the main problems to worry about in abortion are technical. For such people, the only question of interest has to do with the competence and expertise or otherwise of the person performing the abortion and thus the safety with which it can be performed. This, I believe, is a dangerous view. Of course, incompetence is not only a good reason but also an absolutely sufficient reason why someone should not perform an abortion. But competence to perform is not in itself a justifying reason to perform and certainly never a sufficient reason. Several non-technical, non-medical factors are relevant to the issue. It is ironic to worry about the health of the pregnant woman while turning a completely blind eye to the life of the foetus. The view that the foetus is simply a part of the pregnant woman's body, which she should be free to do with as she pleases, denies the foetus not just personhood but humanity. This denial is one impetus for the polar-opposite extreme position that imbues the foetus with autonomy, sacredness and absolute inviolability.

In many countries today a woman can have an abortion on demand without the slightest ado. In other countries, which profess or legislate restrictions, secret abortions are an open secret in a highly lucrative business. But, as Bernard Häring (1972, p.95) has wondered:

> What can be our trust in a doctor who earns his money above all by aborting healthy foetuses instead of healing the sick and to the whole medical profession who seem to approve light-heartedly aborting healthy foetuses instead of healing the sick...?

On the other hand, the view that the foetus possesses absolute inalienable/inviolable rights, which should not under any conceivable circumstances be violated, is also extreme and leads, inevitably, to irresolvable dilemmas. Such a view is clearly implied in the teaching of the Catholic Church through the ages, as is very clearly and unequivocally enunciated in the papal encyclicals: *Casti Connubii* (Pius XI, 1930), *Humani Vitae* (Paul VI, 1968) and reaffirmed, *inter alia*, in the following didactic documents:

(1) "Adress to the Catholic Society of Midwives" (Pius XII, Oct. 29, 1951);
(2) "Discourse to Participants in the Twenty-third National Congress of Italian Catholic Jurists" (Paul VI, Dec. 9, 1972);
(3) "Discourse to those taking part in the 35th General Assembly of the World Medical Association" (John Paul II, Oct. 29, 1983);
(4) "Instruction on Respect for Human Life in its Origin and on the Dignity of Procreation: Replies to Certain Questions of the Day" (Congregation for the Doctrine of the Faith, 1987).

In the teaching of the Catholic Church, human life must be unconditionally respected from the moment of conception, and abortion, for whatever reason, is lumped together with infanticide and murder as "abominable crimes". This position, not only has not changed through the ages, it is even declared **unchangeable**.

Thus the fruit of human generation, from the first moment of its existence, that is to say from the moment the zygote [i.e. the cell produced when the nuclei of the two gamets have fused] has formed, demands the unconditional respect that is morally due to the human being in his bodily and spiritual totality. The human being is to be respected and treated as a person from the moment of conception; and therefore from that same moment his rights as a person must be recognized, among which in the first place is the inviolable right of every innocent human being to life. (*Instruction*, 1987, pp.13-14).

The above two polar opposite positions on abortion are related to two equally extreme attitudes to human sexuality, one of which regards it trivially and light-heartedly as "fun" and the other which shrouds it with solemnity and forbiddeness as the "conjugal act". It should be possible, I believe, to balance these extremes in a morally and psychologically healthy and responsible attitude.

BETWEEN THE EXTREMES

It is clearly extreme and untenable to argue that abortion is wrong under all conceivable circumstances. There are putative circumstances under which abortion would not only be defensible but highly recommendable. One such putative circumstance is the therapeutic case where aborting the foetus could save the pregnant woman's life, whereas refraining from aborting could only result in death of both woman and foetus. To argue in such circumstances that "nature should be allowed to take its course" would be a failure both in proper moral sensibility and ordinary common sense. A good case of non-therapeutic abortion could also be made, on a case by case basis, on behalf of, say, an immature child who gets impregnated, against her own will and better judgement, such as in rape, for instance. But these putative cases of morally justifiable/permissible abortion do not hang on any so-called "every woman's right to an abortion". There is evidently no such right. Every woman may have the right to freely choose to get pregnant or not to get pregnant. But exercising such a right precludes any further claim to a right freely to have an abortion.

A woman who chooses to get pregnant, deliberately decides to bring about another human being into the world in circumstances in which she could have done otherwise by, for example, abstaining from sex or else availing herself of any one of the several methods of contraception. Freely choosing to get pregnant in these circumstances clearly imposes the obligation to respect, at least, the autonomy and otherness of the being thus created, whether or not it is appropriate to consider it a human person or only a human being, whether or not it is sacred or profane. This last point is important because, as Peter Singer (1979) has demonstrated, arguments for or against abortion that harp on the idea of "sacredness of human life", on the one hand, and the idea of "non-personhood", on the other, all run into blind alleys.

It is important not to load arguments for or against abortion on human speciecism because the line between human and non-human is neither sharp nor straight nor very distinct. It is not easy to draw such a line, just as it is not easy to draw a line between human persons and non-persons in terms of such criteria as consciousness, rationality, competence and capacity to make autonomous choices. But why would we want to draw these lines and to what use do we want to put them if we could draw them clearly?

CONCLUSION
As rational creatures, human beings bear moral responsibility for the entire biological world. But, as limited and fallible beings whose knowledge grows only slowly through learning from past errors, humankind will always be faced with moral dilemmas and perplexities. Such dilemmas and perplexities arise particularly in areas not yet covered by the accumulated existing ethical principles, religious injunctions, cultural practices, positive laws and customs. In such cases, rational deliberation is the only means to resolve the issues at stake. Rational deliberation is also the only means by which ethical principles, religious injunctions, cultural practices, laws and customs, which reflect inevitable human limitations and fallibility and are often in disagreement or conflict, can be evaluated, modified and harmonized or changed.

REFERENCES
Congregation for the Doctrine of the Faith (1987): *Instruction on Respect for Human Life in its Origin and on the Dignity of Procreation.* Vatican City: Libreria Editrice Vaticana.
Häring, B. (1972): *Medical Ethics.* England: St. Pauls Publications.

Omoregbe, J. (1979): *Ethics: A Systematic and Historical Study*. London: Global Educational Services.

Singer, P. (1979): Unsanctifying Human Life. In Ladd, J. (ed.), *Ethical Issues Relating to Life and Death*. Oxford: Oxford University Press.

CHAPTER FOUR

AFRICAN BIOETHICS AND SUSTAINABLE DEVELOPMENT

[A version of this chapter is published in World Development: Aid and Foreign Direct Investment 1999/2000, London: Kensington Publications Ltd., in Conjunction with the World Business Council for Sustainable Development (WBCSD), pp. 115-118, under the title "African Bioethics and Globalization"]

INTRODUCTION

Biodiversity, by which I understand the great number and variety of biological species and forms in the world, and agriculture - the deliberate cultivation of any of these species or forms by human beings for human needs - can be said to be made for each other. Biodiversity is to agriculture what concepts are to a thinker. Without concepts the thinker is as handicapped as a farmer without biodiversity. But, once a thinker starts using concepts to make statements, logic becomes indispensable. In the same manner ethics becomes important when an agriculturalist starts using biodiversity in farming. It is, for instance, an ethical issue whether the farmer uses biodiversity sustain-ably, that is, in such a way as not to destroy it, but rather preserve it for future generations, or otherwise. Systematic concern with the sum total of such ethical issues arising from the use of biodiversity might be termed 'bioethics' without much further ado.

However, the term 'bioethics' has been used in more restricted senses in Western discourse. The term is of recent coinage, but that does not imply that what it describes came into being only recently. As a special branch of morality, bioethics is concerned with questions of right and wrong in relation to life or living things generally and, especially within Western discourse, with controversial issues arising from modern Western medicine, biomedical research and attendant technologies. Professor Van Rensselaer Potter who is credited with coinage of the word 'bioethics' intended it as an extension of ethics meant to cover not only medical ethics but environmental and agricultural ethics, in short, 'the application of ethics to all of life' (Potter, 1996, p.2). He justifiably complains that some people have seized upon the term and tried to restrict it to medical ethics. The restrictive sense of 'bioethics' is, however, important because modern Western medicine, biomedical research and attendant technologies, especially genetic technology, have great potential to transform our lives and our world as we have known them, and because of the rapidity, novelty and magnitude of the problems they raise.

Nevertheless, it is also important to keep the less restrictive and more general sense of 'bioethics' always in mind, if for no other reason than that some of the controversies and dilemmas which arise within bioethics in the more restrictive sense are best tackled by considering bioethics in the more general and generic sense. There are peoples and cultures such as those of Africa for whom Western culture and technology mean little but who, nevertheless, can contribute valid and valuable ideas regarding the fundamental issues, questions, dilemmas and controversies arising within Western technology and practice, if these are considered in their stark simplicity, devoid of the complex conceptual baggage and technical jargon that have accrued around them, thanks to various competing vested interests. This is one reason why globalization is an important concept and process, the other being the fact that Western culture is a dominant, domineering and proselytizing culture which has not left any part of the earth untouched. Moreover, cultures, viewed from any initial point of observation, can be considered as forming intersecting concentric circles (Tangwa, 1992, p.142), the widest of which can be taken as delimiting human culture in general.

AFRICAN DIVERSITIES

Increasing global awareness of the importance of biodiversity has led to further awareness that the problems which arise in connection with the preservation and exploitation of the globe's biodiversity, in short, the sustainable use of the earth's biological resources, can only be tackled from a global perspective. The **Convention on Biological Diversity** - which attempts to address *inter alia*, issues of fairness and equity in the global sharing of the benefits arising out of the world's genetic resources - is an example of such attempt at globalization. Moreover, globalization as a process made possible and inevitable by advances in science and technology, especially locomotion and communication technologies, should slowly but surely be turning the world into a 'rainbow global village'.

There is a very simple and straightforward reason why Africa should be greatly interested in this process of globalization. Among the continents of the earth, Africa presents one of the most remarkable diversities - ecologically, biologically, culturally, linguistically and otherwise. Almost any ecological niche, taken at random, anywhere in Africa, presents fascinating possibilities and challenges. In Cameroon, my own country, for instance, with a population of close to 13 million inhabitants, there are over 240 linguistic/cultural groups, and the ecology, the flora and fauna are representative of what is available anywhere else in Africa. Africa also happens to be the continent that, in the past as well as present, has been most exploited by the predatory peoples and nations of the earth. There is at least the chance and hope that globalization might eventually right the wrongs of the past and make possible a fairer world in the future. In any case, certain problems such as 'global warming' whose devastating effects impinge equally on all irrespective of causal responsibility or even knowledge, make globalization simply inescapable.

Variety and coexistence of a plurality of greatly differing entities and elements are the most remarkable attributes of the African continent. African ethical, religious and metaphysical ideas have, over the ages, been shaped and coloured by this background of diverse data. The pre-colonial traditional African metaphysical outlook can be described as eco-bio-communitarian (Tangwa, 1996, p.192.), implying interdependence and peaceful coexistence between earth, plants, animals and humans, by contrast with the Western outlook which could be described as anthropocentric and individualistic. Within the African outlook, human beings are more humble and more cautious, more mistrustful and unsure of human knowledge and capabilities, more conciliatory and respectful of other people, plants, animals, inanimate things, as well as sundry invisible/intangible forces, more timorous of wantonly tampering with nature, in short, more disposed towards an attitude of **live and let live**. African philosophies, belief and ethical systems, both in their latent and forensic forms, provide the supporting structures for this world-view. Within the world-view, the distinction between plants, animals and inanimate things, between the sacred and the profane, matter and spirit, the communal and the individual, is a slim and plastically flexible one. For instance, the belief that humans can transform into animals, plants, such as trees, or forces, such as the wind, is very rampant within this system and has very significant implications for the way and manner nature is approached. In like manner, metaphysical conceptions, ethics, customs, laws and taboos form a single unbroken piece within this African outlook.

WESTERN CONCEPTIONS AND WORLD-VIEW

Talk about "world-views" may today sound rather old-fashioned and anachronistic in the scientific/technological world with its marvels of modernity, but it is impossible for any identifiable human culture or community not to have a world-view, whether it be aware of it or not. The Western world has, perhaps, undergone more rapid and fundamental changes in its metaphysical conceptions, ideologies, moral thought, and way of life than any other human culture, but these changes have not and cannot rid the Western world of a world-view. Generally, it could be quite validly said that 'possessive individualism' correctly describes the outlook and ideology that rules and moves the Western world.

Among the catalysts of change in the Western world, the *Industrial Revolution* of the 18th and 19th centuries has pride of place. The *Industrial Revolution* drew its impetus from the slogan that 'knowledge is power' convertible into commercial value, from the idea that all knowledge is unqualifiedly good, from the belief that nature is, in principle at least, completely knowable and controllable, and from perception of the universe as something which ought to be explored, subdued, dominated and exploited. These ideas and the attitudes they created had their origins and foundation in Judeo-Christianity, but they led, paradoxically, to secularization, desacralization and profanation of everything in the universe - the very antithesis of the Judeo-Christian ethic. The Industrial Revolution and the technologies resulting from it greatly assisted Western imperial nations in their voyages of exploration, discovery, subjugation, colonization, domination and exploitation of other peoples. The cumulative effect of these very important achievements was to infuse in Western culture a spirit of omnivorous discovery, an automatic impulse towards unifying, patenting, monopolizing and commercializing such discoveries and a penchant for spreading and promoting its ideas, vision, convictions and practices under the guise of universal imperatives of rationality and morality which ought to be binding on all and sundry. Today, Western culture is, indisputably, the dominant culture of the world. And, particularly in the domain of science and technology, Western culture is the acknowledged master at whose feet other cultures sit as pupils and apprentices. But this does not mean that others also have to learn and accept all other things from Western culture. Science and technology in themselves have nothing to do with how, for example, people conceive of or worship God, how they marry or bring up children, how they organize their social system or even the **uses** to which they put technology.

The spirit of omnivorous discovery which the **Industrial Revolution** engendered and made possible in Europeans guided them to all parts of the globe where they discovered peoples and cultures so different from theirs that they felt reluctant to qualify them as 'human'. From then on, Europeanization (Westernization) of other peoples and cultures appeared naturally in their eyes as humanization. It is in this way that both altruistic and egoistic motives became mixed and confounded in the relationship between the technologically very advanced Western world, peoples and culture and other (technologically less advanced) worlds, peoples and cultures. Since the Industrial Revolution, technology has been propelled to great heights by Western commerce and the profit motive, by war and the will to dominate, by pure epistemological and scientific curiosity, as well as by the altruistic urge to improve human well-being.

GLOBALISATION, WESTERNISATION AND BIOSECURITY

From the above, some of the socio-cultural and ethical issues involved should already be apparent. I raise some of them schematically and randomly below. But, before doing so, the following general remarks need to be made. It is an indisputable fact that pressures from rapidly increasing populations make African traditional farming methods increasingly obsolete. The globalization of Western biotechnologies are therefore, at least potentially, very important for Africa. But, while efforts at a global perspective of biodiversity and biotechnologies are *prima facie* very good at the intentional level or, at least, on paper, there is, at the practical level of implementation, the real danger that globalization may simply translate into Westernization, for fairly obvious reasons. Is a truly global Bioethics possible in a world of different cultural groups whose respective material situation, level of technological development, powers, ideas, experiences and attitudes differ rather markedly and who, moreover, are not even equally represented in contexts and fora where globalization decisions and projects are discussed and where Western agencies and multinational companies spend their time, money and expertise lobbying for their monopolistic economic interests? One index of the pertinence of this general question is the fact that talk about biodiversity, biotechnology, biosecurity, etc. is today being increasingly matched by talk about biopiracy, biorade, biocolonialism, etc. Biotechnology may be inescapable for all peoples of the world today and, in principle, it does hold revolutionary positive possibilities in such domains as human health, agriculture and animal husbandry; but it is also fraught with possible dangers of abuse or of accidents and the possible consequence of further deepening rather than narrowing the wide gap that polarizes the modern world into rich and poor nations, into haves and have-nots.

SOME SOCIO-CULTURAL AND ETHICAL CONCERNS
SCIENCE AND NATURE

(1) Nature may be a blind force but evolutionary history would seem to indicate that it has within itself mechanisms for correcting its own 'mistakes'. With the emergence of human beings, evolution became a conscious process. And, with human interventions in nature, especially at the level of the genes, evolution may become a consciously directed process. Like in the case of humanly undisturbed nature, human interventions in nature, especially at the basic level of the genes, are liable to mistakes, but human beings, unlike nature, don't have any ready mechanism for systematic correction of such possible mistakes. This worry is particularly relevant in the domain of human beings where the possibility of, say, cloning, in spite of the scientific excitement it arouses, evokes the danger of a catastrophic fall over a fatal threshold. The idea that nature must be understood, mastered, controlled and profitably exploited by human beings, which constitutes the greatest motive force of Western culture, if not somehow tempered, may, sooner or later, turn this worry into alarm. Such ideas are quite alien to some cultures, such as African culture and way of life, which stand in awe of nature towards which there is, moreover, an attitude of great respect, caution and peaceful co-existence. The general African attitude to nature can be said to be bio-friendly. Such attitude, if lent and borrowed in the process of globalization, could help in tempering and balancing the Western outlook.

(2) Scientific knowledge is, by its very nature, gappy and incomplete, based, as it is, on observation, induction and experience, which are no guarantee of the future state of things. There is no absolute guarantee that our future experience will conform to our past or present experiences. We don't know what hidden hazardous properties may lie within the genetically modified organisms (GMOs) or living modified organisms (LMOs) or transgenic creatures that Western biotechnology is creating today which may take us by surprise tomorrow. Genetically engineered foods may take care of the immediate need for abundance and fight against hunger, but they also raise serious problems. First of all, the immediate risks to human health, animal welfare and the environment from the technology itself, especially if it is poorly understood or inadequately controlled, are, unarguably, enormous. Secondly, the long term risks of consuming such foods are not known, which is to say that it remains risky consuming such foods. Furthermore, is a vegetable or a fruit 'improved' with an animal gene still a vegetable or a fruit? Can a committed vegetarian eat such a genetically modified vegetable with a clear conscience? And, if the animal gene in question happens to be a human gene, would it not be a sort of subtle cannibalism to consume such a product? In any case, do consumers not have the right to know exactly what they are consuming in the case of such novel products? The whole idea of transgenic foods has implications which need to be very carefully considered before it becomes standard practice every where.

POTENTIAL BENEFITS AND DANGERS
(3)Genetic engineering and biotechnology, like technology in general, are potentially dangerous to the environment. Pollution and a resurgence of infectious diseases have been frequently mentioned as some of such potential dangers. And, in areas like Africa, where disposal of ordinary urban household refuse is still a big problem in many countries, such potential dangers from novel technologies need to be taken very seriously in advance. In this regard, the movement of transgenic organisms across national borders also needs to be done with great care and caution.
(4) Genetic technology is, no doubt, important and of great potential economic benefit in the domain of animal and crop production, especially of staple or widely consumed foods. It could help to raise the socio-economic strength of poor countries and to reduce the

wide gap between the developed and developing worlds. But this would very much depend on the fairness with which agreements for the transfer and use of the technology are negotiated, on the general willingness to view the world and all its resources, not as the right of the powerful and mighty, but as the common patrimony of all its inhabitants. There is, as yet, nothing to make one believe that such world-view is about to emerge, thanks to on-going globalization. On the contrary, the maneouvres of certain Western multi-national corporations to monopolize and control emerging biotechnologies point in the opposite direction. So does the refusal of some powerful Western countries to fully endorse and associate with some international conventions, such as the Convention on Biological Diversity, which seem evidently in everybody's interest from a global perspective.

(5) There is, therefore, also well-founded fear and worry that the technologically more advanced countries might simply appropriate, for free, the genetic heritage and accumulated knowledges of less advanced communities, the world over, patent these acquisitions, and go ahead to monopolize and commercialize them in a manner that would put their benefits out of the reach of those communities, in a manner similar to what has often happened in the case of other industrial raw materials, in a manner that has become standard practice since the invention of nationalism, imperialism and colonialism. If the regular releases of an organization like **The Rural Advancement Foundation International (RAFI)** are anything to go by, African countries especially need to take this worry very seriously in advance.

TECHNOLOGY AND COMMUNAL VALUES

(6) In many traditional African communities, it was taboo to sell or otherwise commercialize certain things, such as water, housing, fuel wood, the staple food, etc. Such taboos were meant to ensure that commodities which were in no way a luxury but were essential for mere survival, and which had moreover been abundantly provided by God in nature, should be at the disposal of and within the reach of all and sundry. A taboo against the manipulation and commercialization of such things as genes and gametes would have been seen very much in the same light within such a socio-cultural context. But, through the phenomenon of globalization, the industrialized Western world may be globalizing not only its technology but its socio-cultural

values as well, which, taken together, give some cultures, such as African culture, little or no chance of surviving globalization.

(7) Privatization and patenting of genetic information is also highly problematic in societies with a highly communitarian outlook within which knowledge generally and discoveries in particular are not credited to individuals but considered rather as communal property. The communal benefits of biotechnology and bio-prospecting need to be very carefully considered in Africa. It must not be taken for granted that the Western ideology of possessive individualism needs to be part and parcel of technology.

A CAUTIOUS PIECEMEAL APPROACH

(8) Bio-prospecting, and commercialization on the global level, in a world of unequal potentialities and opportunities, as it is today, may end up only short-changing and disenfranchising poor and ignorant but gene-rich communities of what, properly, should be considered their intellectual property rights. For which reason, given the prospect of bio-colonialism and unfair exploitation, it may not be inappropriate if countries which are not yet ready were to consider imposing, for the time being, a moratorium on bio-prospecting within their borders until such a time that they would have thoroughly considered for themselves the ethical issues and socio-economic consequences involved and enacted appropriate legislation to regulate such activities. This, of course, may leave open the danger that such tardy caution might result in their being left behind by those who, throwing all caution to the winds, are ready to jump with both feet into the biotechnological arena. But that is the inescapable nature of a gamble and it could also work the other way.

(9) If we recall that World Bank and IMF engineered structural adjustment programmes (SAPs), which, a few decades ago, were brandished as the ultimate panacea for the prosperity of African countries have, so far, only helped in enriching a handful of individuals while leaving the vast majority poorer than before, it would not be unwise to adopt an attitude of cautious piecemeal approach, always seeking to know first what is really wrong with the state of things as they are before introducing changes aimed at modifying, correcting or improving what is already there. Some peoples and cultures, through the process of globalization, could be running the risk of themselves helping to promote their own exploitation and impoverishment.

A CONCLUDING PARABLE

(10) Let me end with a parable or fable, similar to the biblical story of Adam and Eve in the garden of Eden, which is told with varying details among many African peoples. Originally, the sky, which provided all human needs, was lower down, close to the earth, and within everybody's reach. All anybody needed was to stretch his/her hands and pluck whatever s/he needed from the sky. But one powerful man, a perfectionist, full of the spirit of reckless adventure, which had greatly benefited him in the past, was not satisfied with having to stretch his arms before getting whatever he wanted and therefore tried to look for ways by which whatever he wanted would simply fall from the sky into his ready hands. The sky got annoyed and receded out of everybody's reach for ever. A very similar fable among the Maka of Cameroon can be read in Geschiere, 1997, p.38-39.

REFERENCES

Geschiere, P., (1997): *The Modernity of Witchcraft: Politics and the Occult in Postcolonial Africa.* Charlottesville and London, University Press of Virginia, pp. 311.

Potter, V.R. (1996): 'What does Bioethics Mean?' in *The AG Bioethics Forum*, Vol. 8, No. 1, pp. 2-3.

Tangwa, G. B. (1992): 'African Philosophy: Appraisal of a Recurrent Problematic. Part 2: What is African Philosophy and Who is an African Philosopher?' in *COGITO*, Winter 1992, Vol. 6, No. 3, pp. 138-143.

------------ (1996): 'Bioethics: An African Perspective' in *BIOETHICS*, Vol. 10, No. 3, pp. 183-200.

CHAPTER FIVE

AFRICAN PERSPECTIVES ON BIOMEDICAL AND ENVIRONMENTAL ETHICS[1]

[A version of this chapter is published in A Companion to African Philosophy, edited by Kwasi Wiredu, Oxford, Blackwell Publishing Ltd., 2004, pp. 387-395]

In the domain of morality correct practice without theory is preferable to correct theory without practice.

INTRODUCTION

The last quarter of the out-going century/millennium has witnessed two very important developments - one at the theoretical level and the other at the practical. At the theoretical level, there has been a significant shift of emphasis in the Western world (the dominant/dominating culture of the last two centuries of the millennium) in its intellectual patterns and theorising fashions, from concern with overly speculative abstract theoretical issues to more practical matters. This shift of focus and emphasis has, perhaps, been dictated by an increasing realisation of the very grave dangers posed to the entire planet Earth and its occupants by what could be described as the most successful aspect of Western culture - its science and technology. Practical/Applied Philosophy, Eco-philosophy, Environmentalism, Developmentalism, Medical Ethics, Feminist Ethics, Bioethics etc., are some of the fruits, among many others, of this intellectual shift in focus and emphasis.

At the practical level, there has been the phenomenon of **globalisation** which can be considered as both a descriptive process and as a prescription. As a descriptive process, globalisation has been made possible and inevitable by advances in Western science and technology, especially in loco-motion and communication technologies. This has led to increased contact between the various peoples and cultures that populate the Earth which, today, is aptly described as a 'global village'. As a prescription, globalisation arises from increasing awareness of both the diversity (ecological, biological, cultural, linguistic) and the interdependence of the various parts, peoples and cultures of the world and the consequent simple deduction that the problems, challenges and dangers facing the world as a whole, even if emanating largely from only a small part of it, can effectively best be tackled only from a global perspective.

Human technology in general, and biotechnology (agricultural and human) in particular, have narrowed the gap between the natural and the artificial, between nature and humanity, between 'God's work' and 'work of human hands', to the extent that some have proclaimed God and/or nature dead. Human tinkering with nature, which can be said to have begun with the discovery of agriculture about 10 millennia ago, and which seems both inescapable and unobjectionable, has evolved, thanks to modern technology, into wholesale interventions, epitomised in the engineering and release of novel artificial life-forms into the environment. Such developments have gradually turned perennial moral concern with the physical environment and with medical practice into moral disquiet and even moral alarm. (Beck, 1992; Mckibben,1989, Kassiola, 1990, McLaughlin, 1993).

ECO-BIO-ETHICS

As Frederick Ferré (1994) pointedly remarked at the Nairobi World Conference of Philosophy on 'Philosophy, Humanity and the Environment' (July 21-25, 1991):

By the time organisms are sufficiently artificial to be patentable, it is clear that the relative weights of nature and culture have reversed themselves. Culture is in the driver's seat and nature is hanging on for dear life (literally!) as we hurtle down unexplored roads, with poor visibility, and with uninspected and untried breaks. (pp. 237-238).

The cogency and pungency of this metaphoric remark, made before mammalian cloning became a scientific fact in 1997, is today even more heightened in its appropriateness. Nevertheless, biotechnology also holds a certain justified fascination for human beings because of its potential positive possibilities in such domains as preventive and therapeutic medicine and in agriculture.

In the face of these developments, human ethical sensibilities and responsibilities are urgently called-for. As human beings, we carry the whole weight of moral responsibility and obligations for the world on our shoulders. The claim that humankind is the apex of biological existence, as we know it, has sometimes been dismissed as an arrogant speciecist claim and contested by some human militants for the rights of animals and/or plants. Less disputable, however, is the fact that, while human beings have putative moral responsibilities towards inanimate nature, plants and animals, besides, of course, fellow humans, inanimate nature, plants and animals cannot be considered, without absurdity, as having any reciprocal moral obligations towards humans. Human interventions in nature could plausibly be justified by appeal to this responsibility, although this does not imply that every intervention is justifiable.

For these reasons, the recent focal attention on Eco-ethics, Environmental Ethics, Developmental Ethics, Medical Ethics, Bioethics - all of which can be gathered into one basket labelled *Eco-bio-ethics* - is not only very appropriate but quite timely.

As already indicated, the urgency of the problems, dilemmas and controversies arising in these domains today are largely due to developments in Western technology, especially biotechnology. Their discussion and possible solution, however, cannot be left exclusively to the Western world. Firstly, because of the power and influence of the Western world, its problems are, willy-nilly, also the problems of other parts of the world, especially those which have special links with the Western world through colonisation, imperialism and domination. Secondly, many of these problems or their effects, in any case, are such that they are no respecter of geographical or cultural boundaries. Thirdly, there is no society or culture without some technology of its own, no matter how modest and rudimentary. Ethical issues connected with such technology or from human interventions in nature generally have, therefore, always been present in all human societies at all times. Their philosophical consideration or solution would, moreover, always have elements of universality and timelessness about them.

Thus, peoples and cultures, even those usually described today as technologically 'backward' or 'underdeveloped' could contribute philosophically valid and valuable ideas regarding the fundamental issues, questions, dilemmas and controversies arising from Western technology and technological practices, especially if these latter are considered in their stark simplicity, shorn of what is merely their Western conceptual wrappings and technical jargon. Human experience and knowledge are too vast and complex for any one person, group of persons or culture. What is true in this regard of individual human beings, namely, that, no matter how knowledgeable and self-sufficient you may be, any other human being, taken at random, no matter how ignorant and helpless s/he may appear, knows at least something that you do not know, is even truer of human cultures in general.

AFRICAN PERSPECTIVE

The 'African perspective' that I hope to draw attention to in this short article can be no more than a broad and rather untidy general framework of beliefs, attitudes, dispositions, outlooks and practices, given which, any of the myriad specific ecological or biomedical problems of the contemporary world might be very differently perceived, appreciated and evaluated. I am not out to show that the directions pointed or the possibilities hinted at are the only way forward possible. It is enough that they be recognised and considered as alternative ways forward.

In the course of this essay, if I sometimes say things that sound overly critical of Western culture and the Western world, it is not because I am unaware of the positive values and advantages of Western culture or of its technology in particular. I could write a whole book in praise of the Western world, its culture and technology, but this is not the occasion for it. I may, on the other hand, also appear to be exaggerating and 'romanticising' African culture and its traditional past particularly. But I am also not unaware of some of the short-comings and weaknesses of African culture, traditional and modern. I could equally catalogue these in a whole book, but this is not the appropriate occasion.

Lastly, the generalisations that I may make here are very ordinary generalisations and not the result of any so-called 'study', 'enquiry' or 'research' and should not therefore be taken as claims which preclude all counter-instances under pain of falsity[6]. Nor does this in any way imply that my generalisations be taken as uncontroversial and unchallengeable. When I talk of an African perspective, for instance, I don't mean to imply one which is necessarily true of every African village. I have first hand familiarity with only Nso' culture and way of life. But, well apart from the common elements, features and attitudes of African culture in general, there is no reason to hesitate designating something of the

---

[6] See Tangwa, 1996, pp. 186-187.

Nso' as 'African' (that could even be considered a logical truth), especially in a prescriptive domain like the ethical where prevalence is not an index of justification, value or recommendability.

I use the article 'African' throughout here in the same justifiable way that I use 'Western' without any necessary implication that some differences or exceptions may not be found within what is thus bracketed. I insist on this point because, quite often, people who themselves use such terms as 'the Western world', 'Western civilisation', 'Western liberalism', 'Western democracy', 'Western technology', 'Western Philosophy and Ethical values' etc., have objected to my description of something as 'African' on the ground that they happen to know some part of Africa or people thereof where the description does not apply.

MORALITY AND CULTURE

THE CONCENTRIC CIRCLES OF CULTURE

I consider human cultures as forming intersecting and overlapping concentric circles. (Tangwa, 1992, p. 142-143). The widest of such circles delimits human culture in general. There are attributes and capacities, such as rationality, free purposive action, fallibility, experience (both personal and vicarious) of joy and sadness, pleasure and pain, happiness and unhappiness, egregariousness, mortality etc., which are common to all human beings, irrespective of time and place. These common attributes and capacities give human beings *qua* human beings a certain common frame, as it were, through which they perceive the world and reality and therein find their common distinguishing characteristic *vis-a-vis* other existing creatures. It is what makes all human beings, at all times and places, equal, in spite of their particularising differences which, from another frame of reference, make them rather unequal; they are equal 'in God's eyes', 'from God's point of view', equal the way children are equal before their parent, in spite of their particularising differences which make it possible and necessary for them to bear different names. This equality of human beings has been well captured in the African proverb which states that 'all human blood is equally red and no human blood is redder than any other'. This type of equality can properly be designated moral equality.

If I step inward of this human culture circle, I enter the circle circumscribing African culture. At this level, my cultural circle intersects with that of all other Africans in all parts of Africa and in the diaspora. This conceptual circle is underpinned not only by a certain wide ecological niche but by certain shared historical and spiritual experiences, a grounding metaphysical outlook and a common general attitude to life. If I continue this conceptual journey, it will take me successively through Cameroonian culture in general, Anglophone Cameroonian culture, Bamenda grassland culture, Nso' culture, etc. until I find myself standing at the centre of my own concentric circles, somewhere in the Lum compound of the Ndzenshwai quarter of Shisong village. Every human being perceives the world, inevitably, through the tinting of the concentric circles of his/her own culture. But it takes critical awareness to realise this fact. From this perspective, no culture, in itself, is superior or inferior to any other culture, just as no human being, *qua* human, is superior or inferior to any other. Racists and supremacists cannot be expected to accept this argument.

MORALITY AND RATIONALITY

I think of morality as concerned most generally with questions of right and wrong in human actions and human conduct. The main regulatory mechanism for morality available to human beings, and the only one to which morality itself can be deemed to be subordinate, is rationality, occasionally supported by intuition. Morality is indeed impossible without rationality. For this reason, children below a certain age, mentally unsound adults and non-human animals are not considered as moral **agents** even though they certainly are moral **patients/subjects** (Taylor, 1986, pp. 14-20). The rational and intuitive capacities or faculties, if you prefer, can be deemed to be the same for all human beings, irrespective of culture, time and space. And for that reason, morality can be considered *sub specie aeternitatis*; it remains the same for all times and places, even though human limitations becloud its perception. This does not in any way imply that human rationality itself is eternally fixed and static. It is quite plausible to think that human morality improves along with improvements in human rationality. For example, the moral system within which 'an eye for an eye and a tooth for a tooth' was a notable and generally accepted moral principle or slogan was not entirely irrational or without some justification. But there is no doubt that the succeeding moral system which replaced this *lex talionis* with 'love your neighbour as yourself' was more humane, more rational and more morally appropriate than its predecessor. Today, clear progress in both rationality and morality is being made by attempts to de-metaphorise and to extend our understanding of the 'neighbour' we are to 'love' beyond the familiar fellow next-door, beyond ethic group, tribe, race and nation, beyond humanity, to embrace all sentient creatures (Singer, 1975 and 1993, Chapter 3), and even beyond sentient nature, through all biotic life (Taylor, 1986, pp. 99-168), to species and entire ecosystems (Rolston 111, 1992).

DERIVATIVES OF MORALITY

Morality as such has a wide range of subsets and derivatives, some of which, in roughly descending order of generality, can be listed as follows: Ethics, Customs/Taboos, Professional Codes and Laws. Ethics can be considered as a set of specific general principles by reference to which any concrete human acts, actions, project, etc. can be morally evaluated and judged right or wrong. Such principles are, no doubt, partly a function of time, place, culture, context and particular existential pressures. It appears to me, for instance, that, although a lot can be said in favour of, say, **autonomy**, the insistence of nearly all Western ethical theorists in considering it as an almost self-evidently true cardinal principle of morality is an effect of the Western ideology and practice of **individualism**. We can say that ethical principles are tinted or coloured by these specificities; for which reason, they need to be constantly reviewed, debated and reformulated, in attempt to rid them of these limitations. But it is clear that this task can never be completed, once and for all, inasmuch as every human situation has its own ineradicable egocentric predicaments. The relationship between morality and ethics has been compared to that between thinking and logic or that between religion and theology. (Omoregbe, 1979, p.3). Without thinking there would be nothing like logic, just as without religion there would be no theology. Logic is based on thought, expressed or unexpressed, just as theology is grounded on religion. As a capacity or an act, thinking is the same for all human beings though it may produce slightly different systems of logic, just as one and the same religion may engender different theologies.

Customs and taboos are specific to particular societies or communities. They can be considered as being essentially aids and short-cuts to morality although they also cover the amoral or morally neutral matters. In any case, they are to be evaluated and justified by reference to morality and not vice versa.

Professional codes of behaviour are a subset and derivative of Ethics, and hence, of Morality, that are not only narrow in scope but liable to reflect the egocentric limitations of particular groups of individuals whose main business is not moral deliberation, let alone systematic moral enquiry.

As for Law, it is unarguable that morally relevant positive laws are a derivative and subset of Ethics. Such a putative law or even an entire legal system cannot be adequately justified without reference to morality, but it would be extremely absurd to attempt arguing for, say, the moral permissibility or otherwise of a putative action by pointing out that it is or is not what the law requires. An immoral law, no matter how it may come about, is unjustifiable and we would normally consider it not only right but, in some circumstances, even morally heroic to challenge and defy such a law. There are, of course, amoral or morally neutral laws and therefore not every law is of moral relevance, but an immoral law is absolutely unjustifiable. There are, however, societies in which the law has become so sophisticated and so important that ethics and morality may give the impression of being silent followers of the law in such societies. But, in general terms, while a human society without any positive laws is perfectly conceivable, one without morality would clearly seem to be both a conceptual and practical impossibility. Morality needs no legislating but laws ought to be moral. I would even venture as far as suggesting that the morally more mature or advanced any society is, the less would be the need for morally relevant laws within such a society.

We could go on to discuss the relationship between morality and religion in similar terms. Every religion would seem to have a strong moral component and most religions are, in fact, characteristically obsessed with morality, at least in their teachings and utterances, to the extent that the very mention of morality immediately recalls religion to the mind of most ordinary people. But morality and religion are quite distinct and separable and the latter is to be justified by reference to the former but not the other way round. An immoral religion, if such could be conceived, would be quite unjustifiable no matter what other values it may purport to defend, but an act cannot be morally justified by simply arguing that it is a religious injunction. I would thus consider even religion as, if not entirely a derivative, as, at least, an aid to morality.

AFRICAN DIVERSITIES AND CONCEPTUAL FRAMEWORK

Variety and coexistence of a plurality of greatly differing entities and elements are the most remarkable attributes of the African continent. Africa is one of the richest and most variegated continents on earth: ecologically, geographically, climatically, biologically, historically, culturally, linguistically, natural-resources-wise, etc. Almost any ecological niche, taken at random, anywhere in Africa, presents remarkable and fascinating diversities and variety. In Cameroon, my own country, for instance, with a land surface of only circa 475,000 km^2, and a population of less than 13 million inhabitants, there are over 240 distinct linguistic/cultural groupings which have all experienced German and English or French colonisation and influences, and the ecology, the flora and fauna are truly representative of what is available elsewhere in Africa. Africa also happens to be the continent that, in the past as in the present, has been and continues to be exploited by the powerful predatory peoples and nations of the world.

African ethical, religious and metaphysical ideas have, over the ages, been influenced, shaped and coloured by this background of diverse data. The pre-colonial traditional African metaphysical outlook can be described as eco-bio-communitarian (Tangwa, 1996, p.192.), implying recognition and acceptance of interdependence and peaceful coexistence between earth, plants, animals and humans, by contrast with the Western outlook which could be described as anthropocentric and individualistic. Within the African outlook, human beings tend to be more humble and more cautious, more mistrustful and unsure of human knowledge and capabilities, more conciliatory and respectful of other people, plants, animals, inanimate things, as well as sundry invisible/intangible forces, more timorous of wantonly tampering with nature, in short, more disposed towards an attitude of *live and let live*. African philosophies, belief and ethical systems, both in their latent and forensic forms, provide the supporting structures for this world-view. Within the world-view, the distinction between plants, animals and inanimate things, between the sacred and the profane, matter and spirit, the communal and the individual, is a slim and plastically flexible one. For instance, the belief that humans can transform into animals, plants, such as trees, or forces, such as the wind, is very rampant within this system and has very significant implications for the way and manner nature as a whole and in its various specific manifestations is approached.

In like manner, metaphysical conceptions, ethics, customs, laws and taboos form a single unbroken piece within this African outlook, to the extent that, in the traditional past, it was not usual to separate them, at least from the point of view of the ordinary members of the community. They all covered what ought or ought not to be done, although sanctions for violations varied accordingly, the severest, perhaps, attaching to taboos, whose violation, whether voluntary or not, attracted consequences not only on the violator as an individual but, sometimes, also on his/her extended family or entire community and called for ritual atonement, cleansing and purification.

RESPECTFUL COEXISTENCE

The traditional world-view within which I was born (that of the Nso' of the grassy highlands of Bamenda in Cameroon) recognises the differences between plants, animals, human beings and sub/super-human spirits, but it does not imbue human beings with any mandate or special privilege, God-given or otherwise, to subdue, dominate and exploit the rest of creation, as is the case with the Western outlook. The Nso' attitude towards nature and the rest of creation is that of respectful coexistence, appeasement and containment. Hence the frequent offering of sacrifices to God, to the divine spirits, both benevolent and malevolent, to the departed ancestors and to the sundry invisible and inscrutable forces of nature. The Nso' Year (*Ya' Nso'*) always begins with fertility sacrifices (for plants, animals and humans) officiated by the King (*Fon*) of Nso', the chief High Priest of the Nso' Kingdom, assisted by three ritual concelebrants: *Yewong* (ritual mother of the Kingdom), *Tawong* (ritual father of the Kingdom), and *Fai Ndẓëndẓëv* (chief great councillor of the Kingdom). Through the course of the year, numerous other sacrifices are offered all over Nso' by the custodians of arable land (*ataangvën*), lineage heads (*ataala'*) and *agaashiv* who normally combine the functions of medical doctor, priest, psychiatrist, psychologist and exorcist.

Among the Nso' even something like the treatment of illness is basically a matter of appeasement and containment. The aim is never eradication, elimination, wiping out, or extermination, but rather to coax and plead with the illness to leave its innocent victim alone. Otherwise, there is no attitude of insistence that illness, on its own, should not exist. If illness did not exist, how then would people die? In Nso' conception, death as an inevitable end is not considered a necessarily bad thing, especially when it is a timely and relatively painless one, neither premature nor overdue (Tangwa, 1996, p. 195).

The general point about this African background is that it is consistent with only a cautious and piecemeal use of technology.

WESTERN CONCEPTIONS AND WORLD-VIEW

Talk about world-views may today sound rather old-fashioned and anachronistic in the scientific/technological world with its marvels of modernity, but it is impossible for any identifiable human culture or community not to have a world-view, whether it be aware of it or not, whether it accepts or disputes the fact. The Western world has, perhaps, undergone more rapid and fundamental changes in its metaphysical conceptions, ideologies, moral thought, and way of life than any other human culture, but these changes have not and cannot rid the Western world of a world-view. Generally, it could be quite validly said that 'possessive individualism' correctly describes the outlook and ideology that rules and moves the Western world.

SPIRIT OF OMNIVOROUS DISCOVERY

Among the catalysts of change in the Western world, the **Industrial Revolution** of the 18th and 19th centuries has pride of place. The Industrial Revolution drew its impetus from the slogan that 'knowledge is power' convertible into commercial value, from the idea that all knowledge is unqualifiedly good, from the belief that nature is, in principle at least, completely knowable and controllable, and from perception of the universe as something which ought to be explored, subdued, dominated and exploited. These ideas and the attitudes they created had their origins and foundation in Judaeo-Christianity, but they led, paradoxically, to secularisation, desacralisation and profanation of everything in the universe - the very antithesis of the Judaeo-Christian ethic. The Industrial Revolution and the technologies resulting from it greatly assisted Western imperial nations in their voyages of exploration, discovery, subjugation, colonisation, domination and exploitation of other peoples. The cumulative effect of these very important achievements was to infuse in Western culture a spirit of omnivorous discovery, an automatic impulse towards unifying, patenting monopolising and commercialising such discoveries and a penchant for spreading and promoting its ideas, vision, convictions and practices under the guise of universal imperatives of rationality and morality which ought to be binding on all and sundry. This situation is further exacerbated by the fact that, in virtually all domains of global human concern, the Western world not only has effective monopoly over the discourse, but possesses high persuasive skills which invariably evolve into lobbying when persuasion fails, change into bullying when lobbying proves unsuccessful, and finally transform into threat or use of naked force when all else fails. Today, Western culture is, indisputably, the dominant culture of the world. And, particularly in the domain of science and technology, Western culture is the acknowledged master at whose feet other cultures sit as pupils and apprentices. But this does not mean that others also have to learn and accept all other things from Western culture. Science and technology in themselves have nothing to do with how, for example, people conceive of or worship God, how they marry or bring up children, how they organise their social system and, most importantly, science and technology in themselves have nothing to do with the **uses** to which people put science and technology.

THE WESTERN WORLD AND OTHER WORLDS
The spirit of omnivorous discovery which the **Industrial Revolution** engendered and made possible in Europeans guided them to all parts of the globe where they discovered peoples and cultures so different from theirs that they felt reluctant to qualify them as 'human'. From then on, Europeanisation (Westernisation) of other peoples and cultures appeared naturally in their eyes as humanisation. It is in this way that both altruistic and egoistic motives became mixed and confounded in the relationship between the technologically very advanced Western world, peoples and culture and other (technologically less advanced) worlds, peoples and cultures. Since the Industrial Revolution, technology has been propelled to great heights by Western commerce and the profit motive, by war and the will to dominate, by pure epistemological and scientific curiosity, as well as by the altruistic urge to improve human well-being.

The Western world today is one which admits no taboos or forbidden areas, thereby allowing the spirit of individual acquisition and the profit motive, greatly aided by a highly efficient technology, to rule the world and even the entire solar system. If we continue discussing these issues, it is only because, as Kwasi Wiredu has very aptly pointed out (Wiredu, 1994, pp. 34, 37, 38), technique is different from knowledge while knowledge does not necessarily imply wisdom and all, therefore, stand in need of moral virtue.

The main underlying problem with the Western background and outlook is its epistemological over-confidence, bordering on arrogance and resulting in recklessness. The constant perennial overturning and correction of past 'knowledge' by new 'knowledge' has not led Western scientists and technologists to be more tentative in their claims or more circumspect and cautious in their interventions in nature. On the other hand, and by contrast, the main underlying value of the African world-view and outlook is its epistemological humility and respectful caution, as befits fallible beings.

BIOMEDICINE

The issues in Western biomedical research and practice that have been most discussed recently include, randomly, the following: experimentations on human subjects, cruelty to animals (including using them in experimentation and killing and eating them), animal liberation/rights, informed consent, euthanasia/assisted suicide, artificial prolongation of life, brain death, transplantation of human organs, sale of human organs, cadaveric harvest/banking/sale of human organs/eggs/sperm, surrogate motherhood, artificial insemination, in vitro fertilisation, abortion, acquisition/use of genetic information, creation/release of genetically modified organisms, animal cloning, human cloning, etc. Apart from those of these issues which concern non-humans (plants and animals), the rest could be roughly categorised into three groups: those relating to the beginnings of human life/birth, those connected with the duration of human life/living, and those concerning the end of human life/death.

Apart from the complexification brought into these issues by overt and covert economic/commercial motives, the main underlying ethical issues involved here are connected with the mechanisation of life and its natural processes. This mechanisation process threatens to turn human reproduction, as we have known it, into mere production, and life and death into similar processes depending largely on mere material calculations involving, inter alia, the reading of graphs, charts, balance sheets, insurance policies, patents, costs and the turning on and off of machines. Western culture can be described as a technophile culture although that is not to say that there are no technophobes in the Western world. Without being either technophilic or technophobic, African culture could show Western culture the way back to those natural human values that Western culture has sacrificed to the god of industrialisation/technology and commerce, if, indeed, it ever had them.

HUMAN REPRODUCTION

Let me dwell a bit on human reproduction as an example. There is enormous interest in the Western world on reproductive technologies. This interest on artificial reproductive methods is, however, evidently and curiously out of tune with the increasing lack of interest in natural reproduction and reproductive methods. The accelerated decline in the birth rate in Western countries, thanks to contraception, abortion, family size reduction, homosexuality, celibacy, child sex, etc., is, prima facie, hard to square with the almost feverish interest in artificial methods of reproduction.

In Africa, there would be an equally great interest in artificial methods of reproduction (or assisted reproduction, quite properly in this case) by those unable to procreate through normal natural methods. But this can be explained by pointing to the high value placed by the culture on procreation and children. May it be that what we are witnessing in the Western world is a flight, not from natural procreation, but rather, for some as-yet-to-be-explained reason, from what I would here permit myself to call natural sexuality? For, no matter what can be said in defence or in praise of such practices as celibacy, lesbianism, male homosexuality, oral sex, sex with infants, etc., the fact is that any human community would be heading for extinction, if these became predominant practices in such a community, and the human species would be heading for extinction, if they became predominant practices on a global scale. Is the great interest in reproductive technologies, and particularly asexual reproduction, especially the prospect of human cloning, perhaps, fundamentally related to this flight from normal sexuality, and only secondarily and superficially to the search for 'the perfect baby'? (Chadwick, 1987, pp. 93-135).

In African culture, children are highly valued; so highly valued that procreation is considered the main purpose of marriage. In Africa, marriage for mere companionship is rare, if not completely nonexistent. In Lamnso' we say: "wan dze wan, a dze lim nyuy" (a child is a child, the handiwork of God). A child is always welcome, no matter how it was conceived (excepting through such a taboo as incest - Tangwa, 1996, pp.197-198), no matter how it comes, no matter how it is. Some parents would nearly beat their daughter to death for getting pregnant 'in the house', that is, outside marriage, but, watch them dancing with joy and celebrating when she eventually delivers her child, and you would not believe they are the very parents who nearly killed her for getting pregnant.

HANDICAPPED PERSONS
Within African culture, a child is accepted and loved unconditionally. It is, no doubt, to support this moral imperative that the culture imbues handicapped children with greater value. Within Nso' culture, for example, a handicapped child is considered 'mbuhme' - a special gift - (of God's). Such children are generally believed to possess extraordinary 'depth' and psychic powers and to be liable to be used by God or the spirits as common disguises or as messengers; for which reasons they are approached and treated with great care and respect. However, this by no means implies that anybody would pray to have a handicapped child. On the contrary, the prayer is always that all the children should be healthy and elegant, like the palm tree, robust and strong, like the iroko. If there were a means of selecting the child before birth, surely no one would select a handicapped one. Just as, were it possible to choose one's personal destiny before birth, as in some fables, surely no one could conceivably choose to come into the world as a handicapped. But, once the handicapped child is already there as a fait accompli, it is accepted with unconditional love.

Within this sort of background, there would surely also be great interest in any technologies likely to assist reproduction or the birth of healthy babies in any way, but it would be considered quite horrendous if such assistance is associated with trade and commerce or presented as a matter of seeking a child of a certain 'quality'. (Chadwick, ibid). In Lamnso' it would be quite impossible to convey the ideas that a certain procedure would help a couple to get a child of a better/higher quality. The spontaneous unanswerable question would be: 'How can one child be of a better quality than another?' Quality-talk in the domain of human reproduction is, perhaps, a case of transferring inappropriate concepts and vocabulary from the domain of positive economics and marketing propaganda, under the influence of a calculative moral theory such as utilitarianism.

The saying that a child is a child and its implications are naturally extended to the saying that a human being is a human being. All human beings are equal, although the general profession of this moral tenet by no means implies its acceptance in practice or the willingness of all to face its all its logical implications. Of course, each human being has a host of particularising/individuating physical and non-physical characteristics: Some are taller, others shorter, some are fatter, others slimmer, some healthier than others, some more resistant to infections than others, some more intelligent, others less so, some highly educated, others less so, some have many material possessions, others hardly at all, etc. But, despite rampant quality of life talk, it makes no sense to claim that one human life is of better/higher quality than another. All this not only implies that every human being must be treated as an end in itself (to borrow Kant's very appropriate expression) but also that s/he must accept his/her finitude and limitations, including accepting mortality with calm and dignity. Within Nso' culture, timely death is accepted quite calmly, if not cheerfully.

TECHNOLOGY AND THE ENVIRONMENT
Although the Western world-view can be described as predominantly anthropocentric and individualistic, and contrasted with its African counterpart which I have described as eco-bio-communitarian, these different outlooks have only potential advantages or disadvantages regarding their actual consequences for the Environment. An anthropocentric ethic, even an individualistic one, if it were sufficiently rational, need not necessarily endanger the environment, just as an eco-bio-communal one need not necessarily exclude all dangers to the environment. It must also be noted that in Western discourse, as distinguished from practice, the entire spectrum of attitudes and positions on the environment vis-a vis technology are indeed to be found. (Attflied and Dell, 1996). The important question, then, has little to do with extant environmental ethical theories but is rather a question about which of such theories can be deemed to predominate given the actual uses to which technology has been put and the actual and possible degrading effects of such usage on the environment and which would be most compatible with preservation of the environment.

Some of the most harmful effects of technology that have assumed the status of urgent global hazards are the following: global pollution (of air, water and soil), global warming and consequent erratic/unpredictable changes in global weather systems, massive risks to plants, animals and humans from toxic industrial wastes and from sophisticated weapons (conventional, nuclear, chemical and biological), risks of upsetting nature's ecological balance, risk of accidentally triggering the collapse of the very foundations of life via gene technology, etc. These hazards are urgent for all human beings in all parts of the world, irrespective of their responsibility or lack thereof in their causation, irrespective of whether they are aware of them or not. It would therefore be a great mistake to think that the solution of these problems should be left to the Western world or even just to Western science and technology, to Western scientists and technologists (DeGeorge, 1994, p. 19). Scientific and technological knowledge may be for the specialists, but the moral rightness or otherwise of the manner and uses to which it is put are certainly within everyone's competence to consider.

SELF OWNERSHIP AND COMMERCE

If I were to be asked to suggest one single idea which I consider to have had the most negative all round consequences within Western culture, I would suggest the affirmation 'I own my own body, my own self.' This absurd affirmation, which makes very little sense when translated into Lamnso', is not a trite truism but the very foundation and corner stone of all Western economic systems, capitalist or socialist. This idea, planted in the fertile intellectual soils of the Western world by some of its most influential thinkers, is the channel, in my view, through which the demon entered, took control of and thoroughly permeated the Western system. It is from this seed that the tree germinated, grew, matured and bore such fruits as private property, commercialisation and profit motive, maximisation of economic advantage, etc.

One area at par with science and technology, in which the Western world outclasses all other worlds, is in commerce and commercialisation. Almost any imaginable thing can be commercialised and marketed in the Western world. A culture in which anything can be sold (including money itself!), in which advertisement and promotion can create and raise the demand for almost anything to match its supply , in which you have to pay cash for everything including some of the most imaginative so-called 'services', is certainly a culture that has to be deaf to many things outside the domain of commerce, especially in the domain of morality.

Many of the most heated debates in the Western world, in all domains of discourse, are driven by underlying commercial motives and implications. If this demon could be exorcised from the Western world, Western science and technology would be the best thing that ever happened to humanity since its emergence in evolution. I am aware that I have already slipped into metaphoric and parabolic language, which are abhorred in Western scholarship. But I cannot find a more appropriate mood of expression for my concluding thoughts. One of the consequences of the demon's confiscation and control of the Western system is that poverty is something to be worried about and to be ashamed of, not only within the Western world but, increasingly, all over the world, thanks to the global influence of the Western world.

POVERTY AND SHAME

The culture within which I was born and within which I grew up took great pains both to ensure that no one should die out of sheer poverty and to emphasise that material poverty is nothing to be ashamed of. In Lamnso', my mother language, we have the following common sayings and proverbs:

A si vishong bong kitan (Better to be poor than to steal).
Wir yo' yi kpuh kitan (No one dies of poverty).
Kitan kı yo' dze lii wir (Poverty does not belong with blameworthy actions).
Bong ke ngah kitan (The poor person is better of).

My natal culture took great pains to ensure that the most basic necessities of life were freely available to all and sundry; there were taboos against the sale or commercialisation of certain things such as the staple food, water, housing, fuel wood, etc.; these are things one gave or received, begged for or simply took but which one never bought or sold. Furthermore, it was the general practice that fruit trees such as avocado pears, kolanuts, oranges, mangoes etc., for example, no matter to whom they belonged, were at the disposal of anybody with regard to fruits which had fallen down of their own accord. Anyone could pick such fruits for his/her own use although no one was permitted to climb on the tree or to grab and shake its trunk to cause the fruits to come down. In the late fifties and early sixties , my natal village, Shisong, used to flourish in all the fruits mentioned above and I remember very clearly how large numbers of people with baskets, some coming from as far as fifteen kilometres away, would join us in the very early hours of the morning to pick fruits which had fallen overnight all over the village.

Besides, the culture had a strong levelling tendency and an egalitarian impulse. This did not, of course, prevent some people from getting rich. There were many individuals generally recognised as being rich people. However, the culture had many ways of encouraging rich individuals to redistribute and swap their riches for honorific positions and titles. As Jimoh Omo-Fadaka has very rightly remarked, 'poverty as it is known today was almost unknown in pre-colonial Africa' (Omo-Fadaka, 1990, p. 180).

CONCLUSION

The process of globalisation which today has turned the world into a veritable global village is increasingly bringing both the benefits and hazards of Western technology and especially biotechnology to all other parts of the globe. Some of the most important and critically urgent of the hazards involve human health and the global physical environment. It is my considered view that there can be nothing wrong with technology in itself but only with motivations for its development and the **uses** to which it is put. The motivations for the development of Western technologies and the uses to which they have been put have been driven by overweening epistemological over-confidence/arrogance, anthropocentrism, possessive individualism, economic determinism, war fantasies/nightmares, and the will to possess and dominate the world and other places and peoples. These corollaries of Western science and technology, its heavy packaging, as it were, are conceptually quite separable, discardable and substitutable with different ones, such as those that might be offered from an African perspective, involving epistemological humility, procedural caution, economic communalism and an ecobiocentric **attitude of live and let live**.

In view of the importance, seriousness and urgency of the matter, we should not be cowed or timid to suggest or attempt such substitution because, while Western science and technology, in themselves, may be the best thing that ever happened to humanity, the reverse side of the coin is arguably the lack of any moral progress within Western culture to match its science and technology, in spite of the deafening moral discourse that has always been present in the Western world. If we consider, for instance, the Second World War, as the most spectacular manifestation of this situation in its contradictoriness, it must be admitted that, while there was, perhaps, a brief lull in the **might is right** mentality that culminated in that horrendous experience, in its aftermath, the same attitude and outlook are today still going quite strong and may, sooner than later, in spite of the United Nations Organisation, trigger the annihilation of all biotic life from the earth. The prospect is too serious, too important and too urgent to be scoffed at or lightly dismissed with a wave of the hand.

REFERENCES

Attfield, R. and K. Bell (eds.) (1996): *Values, Conflict and the Environment.* Aldershot/Brookfield USA/Singerpore/Sidney, Ashgate.

Beck, U. (1995): *Ecological Politics in an Age of Risk.* Translated by Amos Weisz. Cambridge: Polity Press.

Chadwick, R. F. (1987): The Perfect Baby: Introduction. In Chadwick, R. F. (ed), *Ethics, Reproduction and Genetic Control.* London and New York: Routledge.

DeGeorge, R. T. (1994): Modern Science, Environmental Ethics and the Anthropocentric Predicament. In Oruka, O. H. (ed), *Philosophy, Humanity and Ecology: Philosophy of Nature and Environmental Ethics.* Nairobi: ACTS Press.

Ferré, F. (1994): Technology, Ethics and the End of Nature. In Oruka, O. H. (ed), *Philosophy, Humanity and Ecology: Philosophy of Nature and Environmental Ethics.* Nairobi: ACTS Press.

Kassiola, J. J. (1990): *The Death of Industrial Civilization.* Albany: State University of New York Press.

McKibben, B. (1989): *The End of Nature.* New York: Random House.

McLaughlin, A. (1993): *Regarding Nature: Industrialism and Deep Ecology.* Albany: State University of New York Press.

Omo-Fadaka, J. (1990): Communalism: The Moral Factor in African Development. In Engel, R. J. and Engel, G. J. (eds), *Ethics of Environment and Development: Global Challenge and International Response*. London: Belhaven Press.

Omoregbe, J. (1979): *Ethics: A Systematic and Historical Study*. London: Global Educational Services.

Rolston 111, Holmes (1992): Challenges in Environmental Ethics'. In Cooper, D. E. and Palmer, J. A. (eds), *The Environment in Question: Ethics and Global Issues*. London and New York: Routledge.

Singer, P. (1975): *Animal Liberation*. New York: Random House.

-------- --- (1993): *Practical Ethics*, Cambridge, Cambridge University Press.

Tangwa, G. (1992): African Philosophy: Appraisal of a Recurrent Problematic. Part 2: What is African Philosophy and Who is an African Philosopher? *COGITO* Pp. 138-143.

----------- (1996): 'Bioethics: An African Perspective', *BIOETHICS*, Vol. 10, No. 3, pp. 183-200.

Taylor, P. W. (1986): *Respect for Nature: A Theory of Environmental Ethics*. Princeton: Princeton University Press.

Wiredu, K. (1994): Philosophy, Humankind and the Environment. In Oruka, O. H. (ed), Philosophy, *Humanity and Ecology: Philosophy of Nature and Environmental Ethics*. Nairobi: Kenya, ACTS Press.

CHAPTER SIX

RIGHTS AND RATIONING IN HEALTH CARE: SOME RANDOM CONSIDERATIONS FROM THE AFRICAN CONTEXT[7]

[This chapter was first presented at the **Colston Symposium on Rights and Rationing in Healthcare,** *organized by Alasdair MacIntyre, The Centre for Ethics in Medicine, University of Bristol, UK, April 15-17, 1999, but has otherwise never been published before now]*

Every society/community possesses enough resources for the healthcare needs of its members.

PRELIMINARY REMARKS

I have attempted to carry out my reflection on the subject of this chapter against the background of a relatively clean slate, though not a complete *tabula rasa*. I have not, for instance, depended on the assumption that health care is a 'right' which seems to be the favorite metaphor in Western discourse of such and similar issues. I have also largely ignored the rather complex and confused syncretic present situation on the ground in African health delivery systems - colonial legacies which need serious urgent rethinking and restructuring. I have taken what I know of health care in the traditional African setting as the paradigm against which I have raised questions, worries and suggestions. I should confess from the onset that I have not directly discussed rights and rationing in health care in this chapter. But I consider the types of issues I have discussed to be more fundamental, to the extent that rights and rationing in health care may not arise if they were properly addressed.

Some of what I say here may possibly sound charmingly naïve and/or conjure the image of a pagan attempting to lead Christian prayers at a pontifical high mass. But I have found no other way of expressing my thoughts, no firm *locus standi* from which to tackle this issue. I feel my conceptual feet dangling in the air outside

---

[7] I thankfully acknowledge the support of the *Alexander von Humboldt-Stiftung* in writing this paper.

the outline of a common paradigm I am aware is shared by nearly everybody within the Western world, but I must make an effort.

My perhaps recklessly bold claim in this chapter is that every society, every community, possesses enough resources for the healthcare *needs* of all its members, if the importance of health (relative to other needs) is correctly appreciated and resources used in a rational and morally sensitive manner.

INTRODUCTION

Against the general background of Western economic theories and rights campaigns and discourse, it would appear almost self-evident that 'health care' qualifies as a 'right' about which ethical problems and/or dilemmas arise in allocating 'scarce resources', both at the macro level of government or society as a whole and at the micro level of the hospital and individual medical/health practitioners. However, my own reflection on this theme within the African context has deflected along the trajectory of first questioning both the assumption that health care resources are scarce and the notion that health care is a right, before trying to sketch a rough picture of what a health care system founded and grounded on traditional African health ideas and practices might look like.

The idea of a 'right' might be considered as a very good metaphor or heuristic device under which many moral issues, especially those concerning natural justice, have been canvassed in the Western world. It is, of course, an important metaphor. But, strictly speaking, outside the realm of positive law, it could be said that no human being has a right to anything. It isn't a matter of right or desert, for who, in the first place, has conferred such a right or how does any human being come by such a right? It may, of course, be said that there are fundamental or God-given rights, but this, again, is just a metaphor, expressing a moral imperative or conviction the background of whose necessity is a situation of inequality, where such putative 'rights' are not fully enjoyed by some people. It may be more useful practically to concentrate on *duties* rather than on *rights*. People could have putative rights they are unaware of and there is nothing morally wrong in this. Also, one does nothing morally wrong by failing to claim or by waiving a 'right' one is aware of. But it is always morally wrong to neglect or to fail to perform a duty.

Each human being finds himself/herself existing in the world as an unchosen *datum* about which it is too late to do anything. This fact should give every human being a great sense of humility before the world in general and before each of the other worldly co-creatures in particular. Such a humble disposition is indispensable for personal as well as public morality. It is such disposition that is at the basis of the cheerful acceptance of both life and death within my natal culture. The acceptance of death, whether cheerfully or otherwise, is very important in view of the human attribute of mortality. It is very ironic that it is within cultures which enjoy a very high life expectancy that the greatest tendency to cling artificially on to life is to be found. Is there here a paradox of 'the longer you live the longer you want to live'? In Lamnso' we say that life and death should be accepted 'with both hands' (*I woo baa*), for which reason a newly born child is always accepted and loved unconditionally no matter **how** it is, and death, especially when it is timely (neither premature nor overdue), celebrated rather than mourned. Life and death are accepted uncomplainingly because they are considered to be, not of human doing but God's prerogatives and it is S/He who knows best. Given such a disposition, the individual may consider his/her moral obligations towards others in terms of 'their rights' but it seems unlikely that such an individual could consider the reciprocal obligations of others towards him/her as exigeable rights without some abandonment of the moral virtue of humility.

Western economic determinism assumes that, because material resources are not unlimited, they are scarce, for which reason human beings, as possessive individuals, compete for them, each trying to maximize his/her advantages over all the others. But, if this economic model based on individualism and the eternal competition it implies and engenders is discounted for a while, it might appear plausible to suggest that there are more than enough resources in the world for everyone's need, though not for everyone's greed, as Mahatma Gandhi is reputed to have put it. If the material resources in the world, finite though they may be, were to be used rationally, unselfishly and with due moral sensitivity, they are certainly more than enough for the needs, including the health needs of all the creatures that populate the earth. Scarcity is a condition that is created by human irrationality, selfishness and moral insensitivity. If these human impediments to material abundance were to be removed, the world would be an earthly paradise of unimaginable superabundance. This thought is certainly idealistic, and 'realism' might lead us to the view that it as idle and unrealistic because ,it is just not going to happen'. As Rosner (1988, p. 21) remarks, 'To argue that funds should be shifted from highway construction or military budgets to health care is really naïve - it is just not going to happen.' But there is nothing particularly wrong with idealism. The actual cannot be what sets the standards of human ambitions, efforts and strivings. In any case, it might be an easier task to strive for a rational and altruistic world of moral sensitivity than to hope to devise a fair, equitable and generally acceptable formula for sharing 'scarce resources' or rationing medical facilities in a world destitute of rationality, altruism and moral sensitivity.

Ordinarily, it is, in fact, quite false to talk of scarcity of healthcare resources, especially within the 'enlightened West', where pharmacy shops and drug stores are stocked to the ceiling with medicines and where medical research organizations and pharmaceutical companies are so spectacularly 'successful'. It would be more appropriate here to talk, not about *scarcity*, but about *surplus* or, more accurately, about the inability of so many human beings to access or to afford healthcare resources and facilities that are, nevertheless, evidently abundantly available.

THE WESTERN SYSTEM

The Western world is so powerful and so influential that the rest of the world is like a powerless captive hostage before its ideas, systems, philosophies, visions, pre- occupations, obsessions, etc. The Western world also has high persuasive skills, ample facilities for canvassing its points of view and great coercive powers when persuasion fails to achieve its ends. Because of this fact, it takes great temerity and even a little effrontery to attempt criticizing any operative Western system or proposing anything that is out of line or tune with Western existential pressures, trends, styles and fashions. Materially, and especially technologically, Western culture is (indisputably) the most successful human culture; psychologically, it is or should be a troubled culture; and, spiritually, it is (probably) a terminally ill culture. For this reason, it is a very good thing that Western culture, which has been a talking but far from a listening culture, seems poised, given the process of globalization, to lend an attentive ear to the other dominated and marginalized cultures of the world. In the global dialogue that may ensue, one of the most contentious issues is going to be the Western conception of a human society as being necessarily a society of free, independent and competing individuals, each pursuing his/her own interests against those of all others. This conception which prioritizes the economic dimension of life, is, no doubt, largely responsible for the material and especially technological success of the Western world, but it is also what has run the world into some of the many urgent problems, dangers and dilemmas it now faces. It is the spirit of (private) individual acquisition and the profit motive, more than anything else, which have driven Western technology to unprecedented heights of success and spread its application in all domains of life. This spirit has led to the commercialization of virtually everything and every imaginable 'service' of which a human being might have need.

The recent spectacular commercial success of the drug *Viagra* recalls to my mind an experience I had in 1989, following the death of my father. We had gone to the compound of my paternal grandmother's father (*tar yiy tar*) at Mbiim Kitiwum village to perform some ritual burial rites for my late father, as custom and tradition demanded. The lineage head, my positional great-grand father, Fai Mbiim Kitiwun, was well into his sixties and, noticing that his youngest wife had just had a baby, I congratulated him and remarked that I had often wondered about the secret of the virility of people like himself and the late King of Nso', Sehm III, alias Mbinkar Mbinglo (1947-1972), who had about 100 wives and who continued fathering healthy children well into his eighties, whereas some young people of nowadays were quite unequal to their solitary monogamous wife. He laughed heartily and reached for his goat skin bag and brought out a piece of the bark of some tree, broke a little piece of it and gave me saying: 'When a child washes his hands very well, he can be let into the secrets of the elders. When you go back, eat this and flush it down with a cup of palm wine and, when you come back, tell me the results.' When I later chewed the piece of bark and drank palm wine, I experienced a spontaneous erection that persisted for nearly 24 hours. Although I was still hoping to wash my hands cleaner still to be let into the secret of the tree itself, the point of this abridged narrative is that it would never have occurred either to my positional paternal grandmother's father or to me, my complete Western-style education notwithstanding, that the bark of this tree, whose secret he took along with him to his grave a few years later, could be commercialized for profit.

Western technology, especially biotechnology, whose success, power and influence cannot be disputed, has created many problems which cannot be solved by technology itself. The solution of some of these problems could possibly come from the direction of cultures which are usually described as technologically backward or underdeveloped, because these problems, in themselves, are not technological problems, even if they arise from or are connected with technology. Such problems are moral problems which, in their stark simplicity, shorn of sophisticated language and technical jargon, separated from all commercial considerations and motives of profit, are within the reach and competence of all human beings, irrespective of state, status or situation in life. They arise because knowledge is different from and superior to technique, wisdom different from and superior to knowledge and virtue different from and superior to both technique, knowledge and wisdom.

THE AFRICAN SYSTEM

The traditional African system, to the extent that it has survived Western colonization, domination and exploitation, or to the extent that it can still be recalled and salvaged from the receding African past, is characterized, *inter alia*, by its epistemological humility and procedural caution, eco-bio-communitarianism (Tangwa, 1996), egalitarianism, leveling tendency and an obsession with moral uprightness (Tangwa, 1997). African metaphysical conceptions, belief systems, philosophies and practices, moreover, are often an untidy assemblage of a wide range of diverse and sometimes apparently contradictory elements, thanks to the importance accorded to the practice of consensus in interpersonal/inter-community affairs and a deep-seated attitude of *live and let live*. If the present globalization and multicultural trend in the world continues along the right trajectory and is successful, the resulting global system is likely to be very similar to the African system in its untidy accommodation of different diverse elements (Jecker, Jonsen and Pearlman, 1997, p. 32).

The African world is also one in which matter and spirit, the sacred and the profane, the individual and the communal, etc. are still largely inseparable, even if plastically distinguishable. By contrast, the Western world is well known for its perennial dichotomies and its partiality for matter over spirit, the profane over the sacred, the individual over the communal, etc. This partiality is, no doubt, partly responsible for the material and especially technological success of the Western world. And, because of its power, wealth, influence and dominance, the entire world is today reaping both the beneficial and harmful effects of this 'success', to varying degrees in different parts of the globe. Nothing has, so far, escaped the Western 'scientific spirit', a spirit of omnivorous discovery without any taboos or limits, a spirit of unrestrained acquisition, study, classification, patenting, privatization, monopolization and commercialization of 'discoveries'. By contrast, the traditional African world is one in which there are still many taboos, including tabooed enquiries and knowledges.

Taboo in the sense I am using it here denotes, not just something that is forbidden, but something whose contravention, whether voluntary or accidental, is liable to bring severe consequences, not only on the agent of such contravention, but also on his/her entire extended family or community as a whole. For example, adultery is a strictly forbidden act within all African communities, as indeed in all other human communities, and usually draws appropriate sanctions. But adultery is not a taboo. Clever individuals in all communities of the world have always been able to commit adultery undetected or without too serious consequences. On the other hand, incest or sex with minors is taboo among most African communities. They are considered as liable to place the entire community/society at great risk and into jeopardy requiring immediate ritual expiation, cleansing and purification.

MEDICINE AND HEALTH CARE IN TRADITIONAL AFRICA
In traditional Africa, the practice of medicine and the arts of diagnosing illness and treatment or healing were considered as a **vocation** rather than a mere **profession** or **technique**. The practitioners of such arts were considered to be specially gifted. Often, their skills and special innate endowments were detectable at birth and sometimes they were acquired through long and dedicated apprenticeship. In Nso' tradition, the birth of specially gifted or extraordinary children, called *won anyuy* (spirit children), among whom the healers and diviners had pride of place, was always accompanied by special signs, and such children underwent elaborate naming and other rituals. Such children are considered to be embodiments of sundry spirits or of disembodied strangers from other lands acting as messengers of God, usually. They are handled with great care ritualistically. Such specially gifted or extraordinary children fall into two broad categories: those whose 'depth' or superhuman endowment is good and benevolent and who are said to possess

kingly depth (sëm vifone), e.g. twins and other sets, children delivered with the umbilical cord around the waist, with two or more placentas, or with clenched fists etc., and those whose 'depth' is considered bad or malevolent (ariim), usually associated with witchcraft and sorcery. These latter include children born with the umbilical cord wrapped around the neck, those who emerge from the womb hands first, those who fail to cry upon emerging from the womb, those born with teeth, etc. The reasons for the elaborate rituals which follow the birth of such children are, in the first case, to determine their actual origin and mission and, in the second case, to convert their malevolence into benevolence or else divest them of their extraordinary endowments through exorcism. They are then, in both cases, also given appropriate names (Mzeka, 1996).

The rise of medical charlatanism and quackery in contemporary Africa is liable to obscure the fact that in traditional Africa practitioners of the medical and healing arts, like many other artists and specialists, normally did not charge any fees for their services, under fear of losing their special natural endowments, although treated patients, as a matter of custom, always voluntarily came back with appropriate gifts and rewards for their healer/doctor. This was possible because the idea of an occupation was not necessarily tied to the idea of earning a living. Nso' traditional society, for instance, was organized in such a way that what one needed for mere survival was at the disposal and within the reach of all and sundry. For example, the King of Nso' was the titular owner of all Nso' land and anybody could obtain land for farming or building simply by requesting for it. This initial equality of opportunity did not, of course, guarantee equality of results, for obvious reasons. That is why, further, it was taboo to sale or otherwise commercialize certain things, such as the staple food, housing, water, fuel-wood, etc. These are things that one gave out or received, begged for or simply took, but which one never bought or sold. Furthermore, by general custom, no one needed any invitation to partake in a meal in any household and it was considered a sign of great impoliteness to decline sharing a meal, no matter how full one already was. It was also taboo to refuse shelter to any stranger who might show up and ask for it. Such taboos were calculated to ensure that no one would die of sheer basic need, no matter how this might have come about. The barest necessities of life, without which no human being can survive are, clearly, food, water and shelter. (In some climates, appropriate clothing would be added to this list, but in most parts of Africa this is not the case). Nso' traditional society was so organized that these bare survival necessities of life would be free and within easy reach of all and sundry. This, of course, would not prevent some individuals, through what ever means, from becoming recognizably rich. But then again, there were also several ways of encouraging very rich people to redistribute their wealth by swapping much of it for honors and titles. It is within such a context and against such a background that the saying that a human being does not die of poverty (*Wir yoh yi kpuh kitan*) in Lamnso' can be understood and appreciated.

In this setting and context, the very highest value was accorded to health. As long as one was healthy, little else mattered, and the attainment of any other objective was within the bounds of the possible. The obsession with health within Nso' culture is seen in the usual common exchange of greetings and enquiries after the welfare of people who are absent.

A tav feyi (You look very healthy).
Eee, m'dze kitavin. A tav sa feyi. (Yes, I am healthy. You also look very healthy).
A tav ben ki bong feyi. (It is very good when people are healthy like this).
Shii (buuh) dze dji wir a dzea kitavin. (The most important thing is to be healthy).
A si dze kitavin ben a kuna tong ka? (Being so healthy, one has nothing really to complain about).
Aaii, tav vi sha vifa vidzem. (Indeed, health is superior to everything else).
Wir ve a lemnin fo bam a dze le? (How are the people you left behind?).
A wuni dze atavkin feyi adzem. (They are all very well/healthy).
Beri nyuy feyi. (Thanks be to God!).

When anybody fell ill in this type of setting, everyone was immediately very concerned. Western-type medical care in Africa has had to adapt to having large crowds around a patient's bed. If the healer needed anything for his/her medical preparations, say, the young shoots of cocoyam plants or immature corn cops, s/he could harvest them from the nearest farm, no matter to whom it belonged, and in spite of the taboo on tampering with immature crops. In other words, all the resources of individuals as well as those of society as a whole were mobilized in the effort to cure the sick, any sick person. The idea of scarcity and dilemmas of allocation of medical resources could not come in because there was no epidemic and the treatment in question was not unreasonably expensive. If an epidemic brook out, the community made the best of the resources at its disposal to tackle the unusual situation, but the idea of preferential treatment in such a situation would have been truly scandalous. The sole qualification for accessing treatment was that one was ill. If some possible treatment were so expensive or so far away that it could not be accessed, then it was as good as no treatment at all and no one would complain of having to die when such a treatment was theoretically possible. In cases of untreatable illnesses, people suffered in silent hope, in cases of terminal illness, they prayed for a quick release by death from needless suffering, and, in any case, they did not seek to teach God or Nature their job.

IMPLICATIONS FOR MODERN HEALTH CARE
The first very important presupposition of health care within the traditional African setting is the equality of all human beings. This human egalitarianism is, of course, also everywhere loudly professed in the modern world, but it does not take much observation to realize that only lip service is being paid to this very important moral principle. The actual world in which we live today is one of great humanly created disparities between both individuals and communities/nations, thanks to the spirit of competition, inequality of opportunities, the principle of *might is right* and the attitude of 'what we have we hold no matter how we came by it'. The second implicit presupposition and one that is particularly relevant for health care concerns is that medicine should be directed primarily at illness and maintenance of health and not towards improving the natural endowments of human beings, creating new capacities, correcting what God/Nature has created or towards the artificial production or prolongation of life. The types of services that have made modern Western medicine prohibitively expensive and unaffordable even for some of its own citizens, and on which the greater percentage of research resources and energies are expended, do not have to do with prevention or treatment of any illnesses but rather with services

aimed at combating nature, such as with plastic surgery, transplant surgery, artificial reproduction and artificial life support, etc.

Having a crooked nose or large breasts or a bald head is not an illness and God/Nature certainly did not intend that a human being should continue his/her life with the help of a machine or with another person's heart or kidney. Every human community has its giants, dwarfs and hunchbacks etc., and there is nothing wrong with such people. If any of them falls ill, s/he would need treatment like any other person. It cannot be the task of any healthcare system to eradicate Nature's anomalies. It is an extremely irresponsible parent who would use resources which could have gone towards the education of his children for planting hair on the bald patches on his head.

From the hand of God/Nature, human beings come in all dimensions, in all shapes and sizes, in all shades and colors, in a way very similar to the biodiversity of the living world in general. It does not appear to me desirable to attempt uniformizing and standardizing human beings according to any putative humanly conceived blueprint of perfection. A world of perfect clones, which, at last, seems within reach of Western biotechnology to bring about, does not at all appear attractive to me and those who are fascinated by its prospect should try to conceptualize its consequences very carefully before bringing it about.

To the extent that such things are legitimate, they can be considered as exotic or luxury medicine to which individuals who insist and can afford them can be permitted to have recourse. But such luxury medicine should be carefully separated from normal medicine, involving primary health care and treatment of illness. If this is done, my contention is that every community/society, no matter how poor, has enough resources for the health care needs of its members. And if the whole world, as a global community, were to act with rationality, altruism and moral sensitivity, there are more than enough resources in the world for the health care needs of all its inhabitants to make the concept of rationing simply meaningless.

It should, however, be sadly recognized that most modern governments prefer spending the collective resources of their own people on such things as arms and munitions, propaganda and prestige projects rather than on health care needs. 'It has been estimated that for the cost of one expensive bomber in the United States, one could provide sufficient dialysis machines to satisfy the needs of every person in the United States who requires dialysis' (Rosner, 1988, 20). Why is it that when it comes to wars (cold and hot) and to weapons (conventional, biological, chemical, and nuclear) there are no dilemmas of scarcity warranting discussion of rationing? Would it not be far better to address such a question which could lead to the elimination of scarcity than to engage in casuistry of fair allocation of scarce resources?

For Africa, at any rate, I believe that, were she to be recompensed for the exploitation and ravages of colonization, and were she to learn useful lessons from the Western world instead of spending all her time and energies striving to imitate her colonial masters, she would have more than enough resources for the health care needs of all her inhabitants. But, of course, realism should not make us forget that this is an unlikely possibility under the present state of affairs. Africa is today even in greater danger of more subtle forms of colonization and exploitation such as biological colonization and exploitation at the level of the genes. However, idealism should not permit us to accept this looming prospect with equanimity.

REFERENCES

Campbell, A., Charlesworth, M., Gillett, G. and Jones, G. (1997): *Medical Ethics*. Melbourne/Oxford/New York: Oxford University Press.

Jecker, N. S., Jonsen, A. R. and Pearlman, R. A. (1997): *Bioethics: An Introduction to the History, Methods, and Practice*. Boston/London/Singapore: Jones and Bartlett Publishers.

Mzeka, N. P. (1996): *Rituals of Initiation in the Western Grassfields of Cameroon*. Bamenda: Kaberry Research Centre.

Rosner, F. (1988): 'The Allocation of Scarce Resources - Rationing of Medical Care', in S. R. Benatar (Ed.), *MEDICAL ETHICS: Proceedings of the 4th UCT Faculty of Medicine Symposium, February 1988*, University of Cape Town Printing Department.

Tharien, A. K. (1995): *Ethical Issues in the Progress of Medical Science and Technology*. Voluntary Health Association of India (VHAI).

Tangwa, G. B. (1996): 'Bioethics: An African Perspective', in *BIOETHICS*, vol. 10, No. 3, pp. 183-200.

--------------- (1997): 'Genetic Information: Questions and Worries from an African Background.' (Paper read at the Conference on *Genetic Information: Acquisition, Access and Control*, University of Central Lancashire, Preston, UK, 1997. To appear in the *Proceedings* of the Conference edited by Alison Thompson and Ruth Chadwick, Plenum Publishers.

CHAPTER SEVEN

MORALITY AND CULTURE: ARE ETHICS CULTURE-DEPENDENT?

[A version of this paper is published in the journal **Turkiye Klinikleri: Journal of Medical Ethics, Law and History,** *Vol. 12, No. 2, pp. 92-97, 2004, and another version in]*

ABSTRACT

In this chapter, it is my contention that cultural diversity is a value akin to biological diversity. As such, it is desirable or at least unobjectionable for a thousand and one cultural flowers to bloom. Moreover, no culture qua culture is either superior or inferior to any other culture. Moral diversity, however, is not a desirable value and universalizability remains the chief identification mark of a genuine moral imperative. Divergence of moral opinion, both within and between cultures is, nevertheless, a palpable fact. Such divergence in my opinion is attributable to human limitations, ego-centrism and fallibility. Moreover, moral divergence over particular issues in no way cancels the broad moral consensus, evident across all human cultures, over fundamental and general moral imperatives. Genuine moral progress at the global level would, no doubt, seem capable of leading to a narrowing in the gaps of moral divergence, although divergence itself may never completely be eliminated. Ethics, therefore, may tend to be culture-dependent but ought not to be culture-dependent; rather should cultures be ethics-dependent, in the sense that every culture or particular aspects thereof, like all other things human, is justifiable only when not in flagrant violation of morality.

MORALITY AND CULTURE: ARE ETHICS CULTURE-DEPENDENT?[8]

Wisdom is scattered in tinny little morsels throughout the world
- African adage -

INTRODUCTION

Culture is basically a way of life of a group of people, underpinned by adaptation to a common environment, similar ways of thinking and acting and doing, similar attitudes and expectations, similar ideas, beliefs and practices, etc. There is a remarkable diversity and variety in the human cultures of the world and in the ecological niches in which cultures flourish. This diversity, an observable fact, is analogous to the equally remarkable diversity of the biological world, of the different biological species that populate the earth. Cultures and sub-cultures are like concentric circles (Tangwa 1992, pp. 142-143) and there is no human being who does not fall within at least more than one such circle, as the nuclear family or, more ideally, the extended family in its African conception, could, in fact, be considered as delimiting the smallest of such cultural circles. Like biological diversity, cultural diversity is thus a datum of our existence with which we may tinker in the hope or with the aim of giving it a particular shape, color or direction. Such tinkering is as liable to achieve satisfactory beneficial results as unbeneficial or harmful ones. For this reason, cultures, like living things, may, over time, flourish or atrophy. But to attempt introducing biological or cultural changes that are too sudden or too drastic is to run the risk of achieving more disastrous than beneficial results.

Unlike culture, morality is grounded on human rationality and common biological nature, and on human basic needs which, being common to all, irrespective of culture, may be considered as defining what it is to be human. For this reason, divergence of moral opinion, both within and across cultures, is a descriptive fact which is a short-falling from the prescriptive ideal. Moral imperatives are necessarily universal. But moral thinking and practices may differ from culture to

---

[8] The original version of this paper was presented at the Europaische Akademie *Spring Conference on Bioethics in a Small World*, Bad Neuenahr-Ahrweiler, Bonn, Germany, 10-12 April, 2003.

culture and even from person to person within the same culture, because of human limitations, including the impossibility of perceiving from more than a single point of view, the impossibility of being an experiential participant of all human existential situations, coupled with human ego-centrism and human fallibility.

NO HUMAN CULTURE IS PERFECT
Human ego-centrism naturally leads individuals to perceive their own culture as **the** culture, but critical observation and reflection can help to correct such mistaken perception. Professor Michael Novak in his book, *The Experience of Nothingness* (1970, p.16) remarks that every culture differs from others according to the 'constellation of myths' which shapes its attention, attitudes and practices. In his view, it is impossible for any one culture to perceive human experience in a universal, direct way.

...each culture selects from the overwhelming experience of being human certain salient particulars. One culture differs from another in the meaning it attaches to various kinds of experience, in its image of the accomplished man, in the stories by which it structures its perceptions. Of course, men are not fully aware that their own values are shaped by myths. Myths are what men in other cultures believe in; in our own culture we deal with reality. In brief, the word "myth" has a different meaning depending upon whether one speaks of other cultures or of one's own. When we speak of others, a myth is a set of stories, images and symbols by which human perceptions, attitudes, values and actions are given shape and significance. When we speak of our own culture, the ordinary sense of reality performs the same function. In order to identify the myths of one's own culture, therefore, it suffices to ask: What constitutes my culture's sense of reality? (Novak, 1970, p.16).

Culture is like congenital tinted spectacles through which we look at reality. We inevitably impose our particular cultural tint on everything we perceive, but critical awareness can lead us to the realization that 'objective reality' is multi-coloured. No human culture or community is perfect although that is not to say that some may not be more advanced or better-off in some respects than others. This would be a matter of critical appraisal. There may be activities/skills at which each culture is 'better' than all the others, but a culture in general cannot be described as being 'superior' or 'inferior' to another on that basis. The French, for example, may be better at wine making or some other such activity than the Germans, but it cannot on that account be said that French culture is superior to German culture. To say that one culture qua culture is 'better' or 'superior' to another culture is like saying that a donkey is better than or superior to a horse. A donkey qua donkey cannot be superior or inferior to a horse qua horse because a donkey is not a horse nor vice versa. The claim that French wine is better than German wine is a meaningful claim, which may be true or false, but the claim that French culture is superior to or better than German culture is a nonsensical claim, equivalent to the claim that a donkey is better than or superior to a horse.

Cultures qua cultures can be said to be equal in the same sense in which human beings are equal, in spite of great differences in their individual and individuating attributes and characteristics. We could qualify such equality as 'moral' equality, not to be confused with other senses of equality. All human cultures are, however, perfectible, because none is perfect; and none can be perfect, given that human beings, the creators of culture, are imperfect beings. Particular cultures or even human culture in general can, however, with time, progress or retrogress in relation to some putative inter-subjective standard of perfection.

The limitations of cultures are directly related to the limitations of human beings who, both as individuals and as communities, are the creators of culture. Human limitations, especially human fallibility, are impossible of complete eradication, in spite of the very strong impulse, present to varying degrees within all individuals and all cultures, to strive for certainty and infallibility under the invincible impulse and optical illusion that they can be achieved. Such an impulse euphemistically may be described as 'the desire to be God'. However, human limitations need not be a hindrance to striving for perfection or to making clearly recognizable moral or cultural progress.

Susan Sherwin (1999, pp. 202-203) has suggested that we consider conflicting moral theories and differing theoretical perspectives as alternative 'frameworks' or 'templates' through which we attempt to perceive and evaluate problems, through which we may gain complementary and overlapping but necessarily partial perspectives, but certainly not definitive exhaustive truths. We can consider cultures in the same light. Cultures are like tinted spectacles through which we view reality, which we thus necessarily perceive as if 'through a glass darkly'. Sherwin (*ibid*, p. 204) further uses the image of 'lenses', which can be readily switched or even layered on top of one another to get a different 'view' of things. I believe that the attempt to 'change', 'switch' or 'superimpose' cultural 'lenses' is very enriching for the individual and salutary for human culture in general. However, western culture, because of its sheer material success and global dominance, its proselytizing character and evangelical impulse, its high sense of self-righteousness and justificationist approach to actions, admittedly and understandably, has greater inertia in experimenting with cultural lens-changing/switching exercises.

MORALITY AND CULTURES

The main difference between morality and culture is that while morality is necessarily universal in its outlook and concerns, every particular culture, as a way of life of a group of people, is inevitably relative and limited, to that particular group or people. Moral rules are different from all other types of rules. They are general, applying to a wide variety of particular cases and instances and are **perceived** as universal and timeless, not as timely or context-bound. Moral rules, injunctions or imperatives may, of course, be expressed in, mingled/mixed with, or reflected in laws, societal customs, cultural practices, taboos, etiquette, etc., but they should not be confused with these other operational structures of society. Morality is based on simple human rationality, not on any specialized knowledge and it is uncompromising in its demands, superceding man-made laws, political expediency, economic considerations and social customs and practices.

A moral reason is always a good and sufficient justification for changing or abolishing a law, political programme, economic project, social custom or practice, but none of these latter can morally be justified by simply claiming that that is what it is, that is, a law, custom, project or programme. Moreover, universalizability is the chief identification mark of a moral judgment or imperative in the sense that, to qualify a statement or judgment as 'moral' is to imply that it is based on considerations other than the particularistic, the self-interested or egoistic, the timely or the expedient. However, morality is not absolute and moral rules are not exception-less. Moral rules are conceived and formulated by human beings and human beings are epistemologically limited and also fallible beings.

KNOWLEDGE AND DANCING MASQUERADES

In my opinion, all human cultures, like all human beings themselves, are **morally** equal, in spite of great differences in their material conditions, power and influence. Individual human beings come from the hand of God/Nature in multifarious shapes, sizes and colors, but, qua human, they are all equal. To use an idea and image popularized by the African novelist, Chinua Achebe, we can consider morality and cultures as dancing masquerades. A dancing masquerade cannot fully and completely be viewed by any single spectator. To have an adequate but necessarily partial view of a dancing masquerade, it is not possible to remain sitting or even standing on the same spot; moving around to change the viewing position and perspective is necessary.

In the global dance of human cultures, Western culture, the proprietor of modern bio and other technologies, has reached out to all other cultures from a firmly seated position, on account of which it has developed a high sense of transcendentalism. It may be in the interest of all of humanity that Western culture should develop the habit of also standing up and moving around a bit, to view the dancing masquerades from different perspectives; or else, it is to be feared that Western culture, its technology and especially biotechnology, if they continue with their present thrust and momentum, to the total exclusion or disregard of the wisdom of other cultures, could easily occasion the death and burial of human culture in general.

There is a little tale from African folklore, related by Ulli Beier (2001, p. 34), a remarkable German, who overcame his cultural ego-centrism and drank deeply from the cultural wisdom of an African people, the Yoruba of south-western Nigeria:

Although Ijapa was cleverer than anybody else on earth, he was so greedy and power-hungry that he wanted to own the entire wisdom of the world. One day he sneaked into heaven and stole the calabash in which Olodumare (God) had locked up all the wisdom. He hung the calabash on his neck and set out on his way home. When he had nearly reached his house in the forest, he came upon a huge tree that had fallen across the path. Three times he tried to climb over the trunk, three times he fell off. He was really surprised, because he had climbed thicker tree trunks before. All this time a little bird had been watching him. Now it laughed aloud and called: "You fool! Don't you notice that the calabash prevents you from climbing over the tree? If you would tie it on your back, instead

of letting it hang from your neck, you would cross that log easily." Then Ijapa
became so ashamed and enraged about his own stupidity that he took the calabash
off his neck and smashed it on the tree trunk. This is how wisdom was scattered
in tiny little morsels throughout the world.

African wisdom forbids any direct attempt at interpreting the
above tale or trying in analytic fashion exhaustively to draw out its
lessons. To do that would be either to show oneself a fool or to take
one's audience for fools, or both. Ulli Beier himself draws one of the
consequences of the above folk tale in the domain of religion for the
different groups of worshipers of different deities (*olorisa*) in the
following terms: "Unlike Christian churches, these groups of *olorisa*
do not compete with each other, nor do they go out to make
converts. It is the *orisa* (deity) himself who selects his devotee. All
orisa acknowledge the fact that **no one** can be in the sole possession
of all truth, nor is there such a thing as a single absolute truth. There
are many **parallel truths** and only the combined wisdom and
understanding of all the cult groups will ensure the harmonious and
peaceful existence of the town." (*ibid*, p. 47). The moral here for the
so-called great world religions, which in their uncompromising rivalry
have sometimes turned parts of the earth into a veritable hell, is too
obvious to require drawing out.

CONCLUSION
Let me conclude by stretching some of the consequences of these
African metaphors and parables to what preoccupies and obsesses all
of us at moment – the war and it aftermath on/in Iraq. All individual
human beings and all individual human cultures dream their dreams.
And dreaming, at both the individual and collective levels, is
harmless, provided there is no possibility or means of translating such
dreams into reality. Recently, I dreamt of grabbing U.S. President,
George Bush, and U.K. Prime Minister, Tony Blair, by the throat in
each of my strong hands and throttling and shaking them like rat
moles, and knocking their heads together, to dissuade them from
going to war in Iraq. That is as far as my pacifist, anti-war obsession,
thoughts and action would go: a harmless dream. But, if there were
the slightest possibility or means of translating such a dream into
reality, it would become a dangerous dream which should perhaps
not be dreamed.

And, talking about the war on/in Iraq, it is necessary, before historical amnesia sets in, to recognize, without any equivocation, that the war had no moral justification. If Hitler had won the Second World War, his victory would not have been devoid of celebratory chanting and dancing all over the world or of some good consequences, such as transforming the world into an orderly earthly paradise, according to some putative Nazi conceptual blueprint. War cannot be justified solely on grounds of its purported good consequences. But, even relying solely on consequences, it is quite hard to accept that the innocent victims of the war in/on Iraq – including those of 'friendly fire', sheer accidents and collateral damage, let alone the enormous physical destruction – were a justifiable price for the elimination of Saddam Hussein or the overthrow of his dictatorial and murderous regime, objectives which certainly could have been achieved at less cost.

The Iraq war was the result of a day-dream of omnipotence, part of whose advertised objective was to turn Iraq into an earthly paradise. But the dream has turned rather into a nightmare, as Iraq evolves into a hell worse than that over which Saddam reigned. Had there been reliance on the collective wisdom of all countries, all cultures, there would have been no war in Iraq. And the United Nations Organization, in spite of its weaknesses and shortcomings, is well-placed as a forum for harnessing the collective wisdom of all countries and all cultures, provided some of its members are not accorded preeminence or permanence on grounds other than their sagacious endowments.

You don't need a club to kill a mosquito; we kill a mosquito with a small clap between the palms of the hands. If a mosquito should perch on the tip of my nose, and, because you love me and hate the mosquito as much as I do, and because you posses an arsenal of hammers, you smash my face with a sledge hammer to kill the hated mosquito, with or without a promise to rebuild it afterwards, you have gravely failed in your rationality. The war on/in Iraq, in spite of its unfolding good or evil consequences, could signify a grievous failure in human rationality, the more so for having been the coldly calculated action of greatest leaders of some of the most self-conscious/self-righteous human cultures.

Persons and cultures with the possibility, capability and means of transforming their dreams into reality need to dream their dreams very carefully. And this is as true in the domain of war as in that of biotechnology or any other.

REFERENCES

Beier, U. (2001): *The Hunter Thinks the Monkey is not Wise… The Monkey is Wise, But He Has His Own Logic.* Wole Ogundele, (ed). Bayreuth: Bayreuth African Studies 59.

Ezenwa-Ohaeto(1997): *Chinua Achebe: A Biography.* Oxford/Bloomington & Indianapolis: James Currey/Indiana University Press.

Novak, M. (1970): *The Experience of Nothingness.* New York: Harper and Row.

Sherwin, S. (1999): 'Foundations, Frameworks, Lenses: The Role of Theories in Bioethics'. In *Bioethics,* Vol. 13, No. 3 / 4, pp. 198-205.

Tangwa, G.B. (1992): 'African Philosophy: Appraisal of a Recurrent Problematic. Part 2: What is African Philosophy and who is an African Philosopher?' In *Cogito,* (Winter) pp. 183-200.

CHAPTER EIGHT

BIOETHICS, BIOTECHNOLOGY AND CULTURE: A VOICE FROM THE MARGINS [9] [10]

*[A version of this chapter is published in **Developing World Bioethics**, Vol. 4, No. 2, pp. 125-138, December 2004]*

One of the most remarkable things about the world in which we all live, localized here on planet Earth, is its biodiversity (the enormous variety of its living forms). Another is its cultural diversity (the enormous variety of its different human cultures). Equally remarkable is the variety, the different forms, heights, and weights, shapes, sizes and complexions with which individual human beings, even within the same culture and locality, come from the hand of God/Nature. I perceive great positive value - if you would permit the emphatic tautology - in this differentiated diversity and variety.

What all human beings have in common, in spite of their rather palpably striking differentiation and differences, is the fact that they are all human beings, equally liable to being, *mutatis mutandis*, rational, self-centered, sociable, fallible, altruistic, equally liable to experiencing sadness/joy, pleasure/pain, equally vulnerable and liable to suffering, equally mortal in the end, in spite of everything else, life-prolonging technologies included. What all human cultures have in common is that they are all creations of human beings, reflecting, on the one hand, human capabilities, goodness, ingenuity, wisdom etc., and, on the other, human limitations, fallibility, frailty, perversity, foolishness etc.

Morality, of which ethics, law, ethos, etc., can be considered as important sub-sets or derivatives, is an essential component of every culture, because no human being, no human society, is perfect,

---

[9] The first draft of this paper was read at the opening session of the annual conference of the Canadian Bioethics Society, Edmonton, Alberta, Canada, on October 28, 1999.

[10] I would like to thank Dr. Ruth Landau of the Hebrew University, Jerusalem, and Dr. Virginia A. Sharpe of the Hastings Center, Garrison, New York, for their useful critical comments on an earlier draft of this paper.

although all are capable of striving towards perfection. No society, no matter how small, can survive and endure, let alone prosper, without morality, without pervasive perennial concern with matters of right and wrong in human conduct and behavior. No human society or culture, I believe, can completely disregard morality and survive for any length of time. The most palpable thing about cultures may be how different they are, each from the others, just as the most palpable thing about individual human beings, even within the same culture, may be how very different they are each from the others. But, just as the color of human blood is the same everywhere, in spite of human differences, it can be said that the identical blood of all human cultures is morality, understood in its simplest conception and function, without which any culture would be truly lifeless, dead. Pervasive perennial concern with morality has given rise to and is reflected in human customs and traditions, in mores, laws and taboos, in group ethos and general etiquette, even though some of these are liable to be equated or conflated and confused with morality proper.

CULTURE AND MORALITY

I am a cultural pluralist. I perceive great value in the remarkable diversity and variety of human cultures, which seems to me remarkably analogous to the biodiversity of the living world, in which I find equal value. I believe that God/Nature had good reasons for not cloning human beings so that each would be an identical copy of all the others, but rather made each according to a unique and unrepeatable formula. The prospect of a world of identical clones, created or manufactured according to any putative formula of perfection, which, at last, seems within the manipulative reach of human bio-technologists, appears to me a rather boring and undesirable one. I am apprehensive of monocultures: human, faunal, floral or agricultural. Let every culture flourish in its own right on its own terms. Every plant growing in the wild, every creature crawling the earth, flying above or swimming beneath it, may have its own inner 'reasons' for being there, its hidden teleological and ecological purposes, or a value in the overall scheme of nature, unknown to human beings. The same may be true of every human culture.

[But] I am a moral Universalist. I believe in the absolute moral equality of all human beings, no matter their particularizing and individuating characteristics, no matter their situation or condition in life, no matter what culture they belong to. I do not believe in arbitrary double standards in morality, in spite of not knowing of any extant moral theory that would be universally accepted without question or one that would explain away, to everybody's satisfaction, divergence of moral opinion. Divergence in moral opinion, whether *inter* or *intra* societies and cultures, moreover, seems to me to be connected with human epistemological limitations and intellectual weaknesses and with human egoism and self-centeredness. In other words, I do not think that we need to be searching for the reasons for moral divergence within morality itself. There is nothing wrong with morality; but there is something wrong with human beings, with human epistemological capacities and capabilities, with prejudice and human perception, with human feelings and desires, with human motivations, emotions and ambitions. The litmus test of authentic moral judgements, for me, is **universalizability**. I believe that every genuinely valid and uncontaminated particular moral judgement is universalisable, although not every such judgement is necessarily absolutely exceptionless. To assume absolute exceptionlessness for any particular moral judgement is to presume a degree of epistemological comprehensiveness not possible with human knowledge.

A DANCING MASQUERADE

Morality may be compared to a dancing masquerade - to borrow an image popularized by the famous African novelist, Chinua Acheb. There is a single masquerade dancing but different people are having slightly different views of it, depending on where each person is positioned, and no one can have a really good and adequate view of it who remains sitting or even standing on the same spot. It is necessary to move around a bit, to change your spectacle position from time to time, to have anything like a near-adequate view of a dancing masquerade. But no matter how hard you try, no one can, from a single chosen position, view the masquerade fully, completely and adequately, because the masquerade itself is not static but dancing all the time. An attempt at a holistic view of reality, culture or morality comes up against an insurmountable egocentric predicament. But, while it would be too much to hope to banish all ambiguity and uncertainty from the realm of moral reflection and discourse, it, nevertheless, seems certainly possible to be able to narrow the gap of moral divergence through disinterested reflective deliberation, such as, I presume, we are having here, as opposed to engaging in 'winning strategies in the moral language game'.

Such disinterested reflective deliberation can start from non-controversial ethical principles such as the so-called **golden rule** (do unto others as you would have them do unto you) from which the venerable German philosopher, Immanuel Kant, derived his famous doctrine of the categorical imperative, which can be simply stated as **'have a good will'** or the Hippocratic *primum non nocere* (above all, do no harm) - the first cardinal tenet of medicine and non-iatrogenic medical practice (Virginia A. Sharpe and Allan I. Faden, 1998). On these two principles, which are, in fact, one single principle, stated in two different ways (positively and negatively), any valid system of morality can be erected and no genuine system of morality can dispense with or reject them. The principles of beneficence, nonmaleficence, autonomy and justice, which have been much discussed in the Western world as the foundational pillars of bioethics, are directly derivable from this single double-sided principle. In my natal language, *Lamnso'*, it is captured in the single idea of *shiliv she jung shi* (literally, a good heart). Human beings may never become perfect or infallible, but human beings can progressively and continuously become more rational and more morally sensitive, because they are infinitely perfectible.

THE QUEST FOR CERTAINTY

The **quest for certainty** has played a very important part in the intellectual and philosophical history of the Western world. But since Rene Descartes, the generally acknowledged 'father of modern Western philosophy', perennial preoccupation with certainty had transformed into a veritable obsession. This obsession, when combined with science and technological progress, with power and the will to dominate, with commerce and the profit motive, like an efficient broom, sweeps everything out of its way, including what we may call 'counter-knowledge' (any knowledge contrary to the dominant paradigm), which I consider to be extremely important in the human context. The obsession with certainty, and the illusion that may be induced of having achieved it in many domains of human concern, is what has given the Western world its spirit of epistemological over-confidence, an *over-sabi* bordering on arrogance, its evangelical and proselytizing impulse, its high sense of self-righteousness that could easily result in heedless recklessness at the level of practice. In the 'Federation of World Cultures (Mazrui Ali, 1976), the Western world is blessed with a big and loud voice, no less than sturdy muscles, has effective monopoly over the discourse in virtually all domains of human concern, and ample skills and facilities for pushing its point of view, thanks to canvassing and promotion, persuasion and lobbying, bullying and coercion. This descriptive assessment, if it has a modicum of validity, calls for great moral circumspection and conscience examination; however, let me not start sounding like an evangelist preacher.

SWITCHING CULTURAL LENSES

I completely agree with Susan Sherwin (1999) when she urges that we should consider conflicting moral theories and differing theoretical perspectives as alternative 'frameworks' or 'templates' through which we attempt to perceive and evaluate problems, through which we may gain complementary and overlapping but necessarily partial perspectives, but certainly not definitive exhaustive truths. We can consider cultures in the same light. Cultures appear to me like overlapping concentric circles, the tiniest of which delimits the smallest community which any particular individual considers himself/herself to belong to, and the widest of which designates human culture in general (1992). The concentric circles of culture are like tinted spectacles through which we view reality, which we thus necessarily perceive as if 'through a glass darkly'. Sherwin (*op cit*: p. 204) also uses the image of 'lenses', which can be readily switched or even layered on top of one another to get a different 'view' of things. I believe that the attempt to 'change', 'switch' or 'superimpose' cultural 'lenses' is very enriching for the individual and salutary for human culture in general. Western culture, because of its sheer material success and global dominance, its proselytizing character and evangelical impulse, its spirit of self-righteous justificationism, admittedly and understandably has greater inertia in attempting such thought experiments.

But let me, for a brief moment, attempt to lend you one such cultural 'lens', by proposing one such thought experiment, for which I apologize in advance should it shock rather than edify you. If we compare Western Christianity with African Traditional Religion, alias Paganism, alias Heathenism, it appears evident that the former, on account of being a well organized and systematized religion, based on a 'revelation' and 'sacred' scripture, with a highly well-trained clergy

~ 96 ~

and preachers, with monumental places of worship and millions of adherents all over the world, is superior to the latter. Paganism is not a revealed religion, it is not an organized religion with a coherent and systematic body of doctrines, it has no divine founder, no sacred scriptures, no prophets or preachers, no clergy, and no churches. Now, as a Westerner and, perhaps, a faithful adherent of one of the great theistic Western religions, has it ever occurred to you that African Traditional Religion, alias Paganism, might be, in spite of the two contrastive sets of credentials above, **morally** superior to, say, Christianity or Islam?

I am privileged to have been born and to have grown up within a framework in which I had both Paganism and Christianity by equal measures. As a member of both religions, the following facts have often occurred to me: My membership of both religions is usually seriously contested only by fellow Christians, who are fiercely intolerant of the suggestion that any other religion could be 'true'; fellow Pagans see nothing wrong in my dual-membership. There are no 'faithful' in Paganism, since there are no dogmas or a decalogue of dos and donots to be faithful to, thereby giving the Paganian the greatest latitude of freedom - the indispensable condition of genuine morality - including the freedom of non-profession of membership, non-attendance of rituals and prayers, and open dalliance or association with 'rival' religions. When the Pagan priest prays, he prays for everybody, present or absent; he prays for universal peace, harmony and prosperity, for fecundity and health for humans, animals and plants. This being the case, and because 'taking a collection' is never part of the prayers, those for whose benefit he prays need not be present, either in body or in spirit. In praying, say, in the face of a human calamity or other distressing happening, the Pagan priest is sometimes heard seriously questioning God, severely chastising and reprimanding the departed ancestors, carefully probing his own conscience for moral infringements willingly or unwittingly committed.

By contrast, Christians in prayer characteristically heap incredible flatteries on God, emphasize their own privileged situation as specially 'chosen' or 'saved', as against 'unrepentant sinners' and the 'eternally damned' and generally pretend a level of personal piety, humility and meekness that is scarcely humanly achievable. Reflecting on all these and many other peculiarities of the two religions, the suggestion that Paganism might, on account of its non-doctrinaire, non-dogmatic, non-proselytizing, non-discriminatory, non-commercial, non-property acquisition/ownership status, be morally superior to Christianity, in spite of its weakness, non-visibility and lack of influence, has often occurred to me. But I am not out to convert anyone to Paganism. With Paganism, it is certainly possible to achieve salvation outside of Paganism. It has never occurred to any Pagan to think that '*extra Paganismus nulla salus*'. Nevertheless, African converts to Christianity and Islam, who frequently fail to find a formula for peaceful coexistence (witness the persistent fratricidal religious wars in, say, Nigeria) need to be reminded that Paganism is the authentic religion of their forebears, even if some people look on it with condescending disdain.

EQUALITY OF CULTURES

No single human culture is privileged with a holistic and comprehensive view of reality. Every culture selects from the overwhelming experience of being human only certain salient particulars according to its contingent situation and motivating agenda of imperatives. There is no culture that possesses monopoly of disinterested objective thinking, while the others are left with, at best, only the ability to narrate their cultural opinions, prejudices, biases and their folkloric tribal myths and legends. For these reasons, no particular culture, in itself, is superior or inferior to any other, just as no particular individual human being, qua human, is superior or inferior to any other, although that is not to say that some may not be better off in some respects than others. Racists and other supremacists cannot, of course, be expected to applaud this argument and it may be difficult for even non-racists and non-supremacists, especially those wearing the coloured spectacles of very dominant or predatory cultures, to fully appreciate or be convinced by it. But there is no human culture that, if approached with an unprejudiced open mind, without any ulterior motives, would not reveal something positively fascinating and rewarding about itself.

There is a European by the name of Ulli Beier, presently living, I believe, in Australia, who approached an African culture - that of the Yoruba of Nigeria - in such an open-minded and unprejudiced manner and 'got a whole life from it', got fascinated to the extent of considering it superior to his own natal culture in its radiant luminosity and expansiveness. While in Bayreuth, Germany, between 1995 and 1996, on a fellowship of the *Alexander von*

Humboldt Stiftung, I, like many other Africans in Bayreuth, used to spend a lot of time with Ulli Beier and his wife, Georgina, (their home at *Iwalewa Haus* served as a sort of sanctuary for homesick Africans in Germany) and it was very evident that they had a deep and authentic appreciation of African culture that some Africans themselves do not even seem to have.

The authenticity of Beier's appreciation of African (Yoruba) culture stems from the fact that he did not go to Africa, in the first place, with the attitude of the colonialist, the exploiter, the missionary or the researcher. His main reason for going to Africa was to get away from Europe and the traumas of the Second World War. I have come across few non-Africans who, like Ulli Beier, understand so thoroughly the downward trajectory that African cultures and socio-economic systems have taken in the last few decades or who exude such genuine distress over this state of affairs. If some colonialists, missionaries and anthropologists (and some there certainly are) can be bracketed with the likes of Ulli Beier, it is only to the extent that they have been able to transcend the limitations and impediments imposed by their own initial intentions, attitudes and prejudices. Within the domain of culture, the guiding principle should be that captured in the injunction *'When in Rome do as the Romans do'.* But, of course, if you are in Rome as a colonial master, then you would not want to do as the Romans do but rather to get the Romans do as you do or want.

SOME TOPICAL EXAMPLES

Let me now briefly try to anchor some of these generalities on a few concrete topical examples. Some of the recently most widely and hotly debated issues in Western bioethics, with an intercultural dimension, have been so-called Female Genital Mutilation (FGM) and placebo-controlled drug trials[11]. In mentioning these issues here, I am not so much trying to join the ethical debate on these particular issues as attempting to draw attention to what I consider to be the possible egocentric limitations and conditions, the cultural prejudices

---

[11] See *Bioethics,* Vol. 12, No. 4, October 1998. See also Sandra D. Lane and Robert A. Rubinstein, (1996): 'Judging the Other: Responding to Traditional Female Genital Surgeries'. In *Hastings Report,* Vol. 26, No. 3, May-June, Pp. 31-40.

and biases, that have led to the debate and which may have made the debate on each issue seem to be a more serious and more irresolvable controversy than might otherwise be the case. I can best put my point in the proverb that *'those who cherish eating turkey may not be aware that, in all essentials, it is no different from chicken'*. Or, if you prefer a more familiar figure of speech, *'the grass across your own fence always appears quite different from that within the fence'*, especially if your fence is a concrete wall. (In my village we often build our fences with *kwarakwara* - made from the soft but tenacious and pliable inner stuff of the raffia bamboo - because, while *kwarakwara* is more resistant than, say, grass mat, it is flexible enough to permit passage, if you really want to insist. African (Nso') culture, by comparison to other cultures, could figuratively be termed a *kwarakwara* culture, where the others are more or less **cast iron** cultures).

I don't know whether or not it is ethically correct to use placebos in clinical drug trials. I have not yet given enough thought to the issue, although, in general, I would tend towards the opinion that it is unethical and unnecessary to seek to gain any type of knowledge by exploiting other human beings, with or without their informed consent. But, if any placebo-controlled experiment is correctly considered unethical in the USA, then it cannot be ethically right in Africa or anywhere else, or, if indeed it is ethically right in Africa or anywhere else, then it could not have been ethically wrong in the USA or anywhere else. To argue that the rightness or wrongness of such a case depends on 'context-sensitivity', involving socio-cultural, economic and political conditions is clearly to abandon the moral rationale of universalisability, to introduce intolerable relativism into morality, to subject morality to prudential conditions/calculations and to try to use sophistry and equivocation in the service of pragmatic ends and objectives. To argue that a drug trial which cannot be carried out in the Western/first/developed world **on ethical grounds** can be carried out in Africa or any other part of the so-called third/underdeveloped world, on the grounds that those on whom the tests are carried out are those facing epidemic scale of infections, are too poor to buy expensive drugs and stand to benefit the most from the result of such trials, is really to abandon all moral sensitivity. It is a remarkable fact that such a pseudo-moral argument is coming from people within a culture whose moral foundations and underpinnings include the idea of the son of God himself voluntarily assuming human form and altruistically accepting an ignominious death on the cross for the salvation of the whole world.

As for female circumcision, which in the Western world has been given the value-laden and morally condemnatory name 'female genital mutilation' (FGM), it certainly is a cultural practice whose abolition should be encouraged in all appropriate ways. However, the attitude of Western campaigners against female circumcision, while it may not be devoid of altruism, is clearly mainly motivated by cultural bias for the following reasons, *inter alia*: (1) Female circumcision is rightly bracketable with male circumcision, but this is not usually done, probably because important segments of Western society practice the latter unlike the former which is unheard of within Western culture. It should be noted that female circumcision is of many types and degrees of severity as a surgery, some of which are less and others more drastic than male circumcision. (2) Both male and female circumcision share close affinity with some other practices which are prevalent and, more or less, popular within Western culture, such as cosmetic or plastic surgeries, body-piercing, tattooing, etc., over which anti-FGM campaigners are usually completely silent. (3) There is no strictly moral argument against female circumcision, alias FGM, which would not be equally applicable to male circumcision and cosmetic plastic surgery, etc.

I am the author of one such attempted argument[12], the gist of which I can briefly summarize here as follows: Except for indisputably curative therapeutic reasons, circumcision (male and female), and other types of body modification surgeries, is clearly ethically wrong, if carried out without the explicit solicitation and fully mature and well-informed consent of the person on whom it is performed, because it violates bodily integrity, autonomy and self-determination. On no account, therefore, should any of these surgeries be carried out on an infant, child or other morally incompetent human being, for non-therapeutic reasons.

---

[12] See Tangwa, G. B. (1999): 'Circumcision: An African Point of View'. In Denniston et al, (eds), *Male and Female Circumcision: Medical, Legal, and Ethical Considerations in Pediatric Practice*. New York/Boston/Dordrecht/London/Moscow: Kluwar Academic/Plenum Publishers. Pp. 183-193.

BIOTECHNOLOGY

In the light of the foregoing, it is possible also to take a general but clear position on some contemporary debates related to bioethics and biotechnology. One of the central issues debated during the annual conference of the Canadian Bioethics Society, Edmonton, Alberta, Canada, 28-31, October 1999 was the question as to whether bioethics can be local while biotechnology is evidently global.

Bioethics cannot be a local affair, or, maybe, one should say **can but should not** be local. The mark of a genuine ethical judgement is its universalisability and its appearance of being spatially and timelessly valid. An ethical system that is exclusive or discriminatory in any other way, is, *ipso facto*, morally flawed. As for biotechnology, it is, in fact, not global and is free to be as local as it wants to be. In general, there is nothing wrong with technology, as such. In itself, it is morally neutral, neither right nor wrong. It is an important non-moral value, connected with human ingenuity and achievements. But, the **uses** to which any technology is put, is a moral issue. For example, developing, let's say, an infectious contraceptive is a technological affair, it is a local and localizable affair, even a personal/individual affair (check the patent owner), but the 'exploding population' amongst which such a putative contraceptive is released or unleashed, is an ethical matter.

Again, isolating, say, the Ebola virus, is a biotechnological affair, a local affair, which the locality can even leave to its scientists and technologists, its experts and specialists, and their professional competence and consciences. But, attempting to manufacture a biological weapon out of the Ebola virus, is an ethical matter, a global affair of no mean magnitude, whose rightness or wrongness requires no particular expertise to determine; while using such a weapon anywhere, is an even more serious global ethical issue.

So, given the potential dangers of biotechnology, which are now beginning to be placed side by side with its purported benefits, it may be advisable for a brief moratorium to be placed on certain forms of it while the ethics of their usage are discussed or, at least, for such technologies to be restricted to their places of birth and localities, so that innocent people in other places, who have never developed and never benefited from biotechnology may not run the risks of bearing the brunt of it possible unethical uses or even accidental unintended failures.

CONCLUSION

The above musings may seem too general and too diffuse to have any practical relevance in an era of fierce bioethical debates and controversies on specific issues. But it appears to me that 'the devil' is not so much in the details as in some general considerations, assumptions and presumptions that are liable to be taken completely for granted. It may be quite hard, for instance, for many contemporary bioethicists to appreciate the fact that the general substitution of the terms 'consumer' and 'provider' for 'patient' and 'doctor', respectively, in Western discourse is liable to be, more or less, profoundly shocking to some people of non-Western cultures for reasons that are fundamentally relevant to some of the debates.

We are in the era of biotechnology. Nowadays we hear increasing talk, *inter alia,* about sale of eggs. One of the provocative questions in the general theme of the 1999 annual conference of the Canadian Bioethics Society was: "Who decides if eggs are for sale?" In Ndenshwai-Shisong, where I was born and where I grew up, everybody would confidently answer: "The owner of the hen!" Until very recently, everyone in my natal village knew not only the compound to which every cock and every hen belonged, but also the actual owner of each one of them. And no matter where a hen chose to lay her egg, it would always be taken to the owner, who then decided to either sell it, donate it to the Catholic priests or Reverend sisters across the river, or fry it for the children playing in the courtyard. But these are not the eggs whose sale is in question here. By some stretch of language and the imagination, we are here talking about human ova! God forbid! The corruptive influence of commerce and the profit motive should not be allowed to invade all facets of human life. In reproduction, humans are privileged to share in one of God's/Nature's highest prerogatives. This function, and allied processes, is so far beyond a price that it should be kept simple and clean and sacred, and protected from the well-known dangers connected with speculative trading and commerce.

Human scientific and technological knowledge has advanced to the point where scientists and technologists are able to play real games with God/Nature, manipulating the building blocks of living things at will. It is a dangerous game, its purported anticipated benefits notwithstanding, in which they are being encouraged, aided and abated, supported and funded by powerful industries and corporations, for motives of profit. Before the combined might of science, technology and international corporate commerce, ordinary people of all cultures, all localities, all over the world, are quite helpless and have only their right of token protest and boycott as a defensive weapon.

The pharmaceutical industry and its global network have put drugs and health out of the reach of ordinary poor people the world over. They are in an unholy alliance with the biotechnologists and big multinational corporations, trying to take monopolistic control of everything in the world. That is why drugs are becoming ever so scarce and unaffordable, in spite of being abundantly available, to the extent that so many people are dying daily from an upsurge of known and unknown diseases. In Ndzenshwai-Shisong, when I was growing up scarcely four decades ago, there were no pharmacies, but no sick person ever went without medicine or treatment. Even enhancing drugs were freely available. The elders would quietly pass around among themselves their own version of, say, *viagra*, in the same manner they passed around snuff and kolanuts, without letting the non-elders into their secrets. I witnessed my very first death from a cause other than old age when I was already well into my teens. Today, in Shisong, it would be hard to find an infant who has never seen a corpse.

Commerce and the maximization of profit motive constitute, in my well-considered view, the greatest antidote to moral sensibility and sensitivity, generally, and to moral deliberation, discussion and practice (domestic and cross-cultural), particularly. I will, however, not attempt to prove this conclusion, knowing fully well what experts, in the art of disproof, some people are.

REFERENCES

Mazrui, A. (1976): *A World Federation of Cultures*. New York: The Free Press.

Novak, M. (1970): *The Experience of Nothingness*. New York: Harper and Row.

Sherwin, S. (1999): 'Foundations, Frameworks, Lenses: The Role of Theories in Bioethics'. In *Bioethics*, Vol. 13, No.3/4 1999, pp. 202-203.

Tangwa, G.B. (1992): 'African Philosophy: Appraisal of a Recurrent Problematic. Part 2: What is African Philosophy and Who is an African Philosopher?' In *Cogito*, (Winter), 1992, pp. 142-143.

Virginia A. S. and I. F. Alan (1998): *Medical Harm: Historical, Conceptual, and Ethical Dimensions of Iatrogenic Illness.* Cambridge: Cambridge University Press. P. 81.

CHAPTER NINE

CIRCUMCISION:
AN AFRICAN POINT OF VIEW

[This chapter was first published in **MALE AND FEMALE**
CIRCUMCISION: Medical, Legal, and Ethical Considerations in
Pediatric Practice, *edited by George Denniston et al., New York: Kluwer*
Academic/Plenum Publishers, pp. 183-193, 1999]

ABSTRACT

Circumcision, by which I mean any surgical intervention on the genitals of a human being for cultural, religious or purely secular and profane reasons, has recently become a highly controversial issue reminiscent of such other issues as the abortion debate. Pro-circumcisionists have marshalled as many arguments in its favor as anti-circumcisionists have marshalled against it. Quite interestingly, both sides have used science and the work of eminent scientists to support their respective positions, making it evident that science can be a double-edged sword which lends itself readily as an alibi for strongly held preferences or cultural biases.

In this paper, I discuss circumcision as a **rite de passage** within an African culture- that of the Nso' of the north-western grasslands of Cameroon. I then attempt to provide what I consider cross-cultural arguments **against** circumcision without prior informed consent, especially routine infant circumcision. But then, I also argue **for** the availability, in principle, of circumcision under the best medical conditions possible, for well-informed adults who, for any reason, freely insist on undergoing it.

CIRCUMCISION: AN AFRICAN POINT OF VIEW

Those who eat turkey on 'Thanksgiving Day' should think carefully before condemning those who eat porcupine on similar occasions.

PREAMBLE

Before 1996, I was not very aware that the practice of circumcision raises serious medical and ethical problems. I had, of course, especially from 1994, increasingly been aware of the great Western campaign against 'female sexual mutilations', a campaign which looked like an off-shoot of the feminist movement, and which, perhaps because of my own peculiar cultural background, I considered quite appropriate and timely, even if somewhat a little exaggerated. I had heard stories about how, in the USA, some African women, in danger of being deported as illegal immigrants, had successfully used the 'female genital mutilation card' to avert the danger of deportation by claiming that they ran the risk of being forcibly circumcised if they returned to their mother-land. Throughout this paper, I have preferred using the, admittedly inadequate, term 'circumcision' which is more appropriate when the practice is looked at from within a practising culture, rather than the value-laden, morally condemnatory term 'sexual/genital mutilation' which is more current in Western discourse and literature on the subject.

I had first become aware of the existence of the practice of female circumcision during my University student days (1974-1984) in Nigeria where the practice is common among some indigenous groups. I considered it then an extremely strange practice, not knowing at the time that it also existed in some parts of Cameroon. When I tried to enquire as to why anyone would do such a 'senseless thing' as 'circumcising a female', I gathered that it was believed to reduce promiscuity and to facilitate childbirth.

ONE PERSON'S SNAIL IS ANOTHER'S TERMITE

My experience with snails (which Nigerians fondly call 'congo meat') prevented me from passing severe critical judgement at the time on this practice and the purported reasons in its favor. In effect, I had summoned the courage to taste snails (taboo among my own people, the Nso') for the very first time in Binin City, Nigeria, sometime in

1975. I was pleasantly surprised that snails, towards which I felt a certain cultural horror, in fact, tasted so good. While back home on holidays, I related my experience with snails in Nigeria to my family members, and I could see everyone instinctively recoiling away from me as I told the story, until I added that I had met people who were shocked and horrified when I told them that in Nso' we greatly cherish termites, crickets and certain grasshoppers as delicacies.

MYTH AND REALITY

Professor J. O. Sodipo, who supervised my M. A. at the University of Ife, Nigeria, once drew my attention to certain very pertinent remarks of Professor Michael Novak in his book, *The Experience of Nothingness* (1970, p.16). According to Novak, every culture differs from others according to the 'constellation of myths' which shapes its attention, attitudes and practices. In his view, it is impossible for any one culture to perceive human experience in a universal, direct way. It is worth quoting him verbatim at some length:

...each culture selects from the overwhelming experience of being human certain salient particulars. One culture differs from another in the meaning it attaches to various kinds of experience, in its image of the accomplished man, in the stories by which it structures its perceptions. Of course, men are not fully aware that their own values are shaped by myths. Myths are what men in other cultures believe in; in our own culture we deal with reality. In brief, the word "myth" has a different meaning depending upon whether one speaks of other cultures or of one's own. When we speak of others, a myth is a set of stories, images and symbols by which human perceptions, attitudes, values and actions are given shape and significance. When we speak of our own culture, the ordinary sense of reality performs the same function. In order to identify the myths of one's own culture, therefore, it suffices to ask: What constitutes my culture's sense of reality? (Novak, 1970, p.16).

Some people may rush to derive cultural and moral relativity from such remarks, but I don't think that the salient point here implies any relativism. The chief identification mark of morality is universalisability and, although every particular culture is inevitably relative, there are cultural universals (Wiredu, 1990).

THE LIMITATIONS OF CULTURES

Culture is like congenital tinted spectacles through which we look at reality. We inevitably impose our particular cultural tint on everything we perceive, but critical awareness can lead us to the realization that 'objective reality' is multi-coloured. No human culture or community is perfect although that is not to say that some may not be superior or, at least, better off, in some respects than others. This would be a matter of critical appraisal. All human cultures are, however, perfectible. Particular cultures or even human culture in general can, with time, progress or retrogress in relation to some putative inter-subjective standard of perfection.

The limitations of cultures are directly related to the limitations of human beings who, both as individuals and as collectivities, are the creators of culture. Human limitations, especially fallibility, are impossible of complete eradication, in spite of the very strong impulse, especially within certain cultures, such as western culture, to search for certainty and infallibility within the framework of an implicit incorrigible belief in determinism and consequent predictability. Such an impulse, which can euphemistically and humorously be described as 'the desire to be God', not infrequently induces the illusion of having achieved its objective.

CIRCUMCISION IN NSO' CULTURE

As elsewhere in Africa, circumcision among the Nso' of the grassy highlands of Bamenda in Cameroon is primarily an initiation ritual, a **rite de passage** towards adulthood. The Nso' are like the Jews in one respect: they circumcise their male but not their female children. It is not clear why the Nso', unlike many other African groups, never thought of female circumcision, but they make fun of the very idea in the same way that they ridicule, say, the practice of bride-price as 'selling a daughter like a goat'. Unlike the Jews, however, the Nso' do not circumcise as a matter of any divine command nor even from any other religious reason, except in so far as the very idea of ritual might connote and evoke a certain religious feeling. Nso' traditional religion, like African Traditional Religion in general, is, moreover, neither a doctrinaire nor a proselytizing religion. Circumcision among the Nso' is a purely secular and profane ritual of initiation into manhood.

Rituals are very important in Nso' culture and the most important stages in life are always marked by appropriate rituals. For more on rituals in Nso' culture, see my 'African and Western Conceptions of and Approaches to Birth and Death' - paper read during the Third World Congress of Bioethics, San Francisco, USA, 1996 (unpublished).

The term circumcision may be considered a euphemism and the Nso' have two further euphemisms for this euphemism. They call it *nangsin* which literally means to fix, repair or correct, but which also connotes prepare or arrange. They also call it, more obscurely, *sang mbe* which literally means to mark or scarify the shoulders, but which also connotes to toughen, fortify or strengthen.

Mzeka P., (1996, pp. 41-42) has succinctly described the circumcision ritual in Nso' as follows:

... *At the wanle nsum (adolescent) age, boys of a patrilineage were brought together to the compound of the lineage head for circumcision. Before this was done, the 'owner of the spider' should have been consulted to select, using his divination skill, a neighboring expert in circumcision from among those that existed. The one chosen by the diviner was invited.*

The days of circumcision were either nggoylum *or* kilovey *both Nso' days of rest, the so-called 'country Sundays'. The day before the chosen day, mothers of the candidate provided mashed cocoyams (colocasia antiguorum) with egusi soup to be eaten by the boys immediately after circumcision. Their fathers supplied a fowl each, some salt and oil while the lineage head provided a calabash of very fresh raffia wine. Before circumcision, each boy gave the person to carry out the exercise a small amount of money said to be for the man's* kem *or circumcision knife. The fowl was the circumcision fee, while some of the wine was mixed with salt and used for the treatment of the wound and the rest was drunk after the exercise.*

An enclosure was built in the compound of the lineage head and a small pit dug in the centre of the enclosure. The naked boys were taken in one at a time. When in, each boy lay on his back with his penis over the pit and the legs spread out. A small board was placed on the boy's chest to screen his eyes from what was being done. The man performing the circumcision tied off the prepuce with a fibre string and then slashed the foreskin off with his knife. The blood flowed into the pit and the shingyong she wan, *the child's point, i.e. the prepuce, was also dropped into the pit which was eventually filled up and sealed after all had taken their turns. The foreskin was cut in a single stroke and after each slash, fresh wine mixed with salt was taken into the mouth and flushed on the wound.*

After the exercise, the boys went to live in the faay, *the house of the head of the patrilineage for seven days. During the one-week stay they were cared for by their fathers who early each morning came to wash and treat the wounds. Their mothers sent in food but like the rest of the women were not allowed to see them. It was held that if a woman saw the wound, it would never heal. Whenever a boy went out of the* faay, *he carried a small board made of raffia bamboo piths to cover his wound from being seen by any woman who could accidentally be passing by.*

As Mzeka remarks, loc. cit. (p.43), the above description is essentially only a relic of what in pre-colonial times must have been a much more elaborate ritual encompassing several other 'activities pertaining to the acquisition of skills and the inculcation of a communal manhood psyche' as part of the normal education of boys. In chapter 8 of his autobiographical novel *The African Child*, Camara Laye (1959) gives a moving and more detailed description of a similar ritual among the Malinke of Guinea. Another very interesting account can be read in Turner (1971).

RATIONALIZATIONS FOR CIRCUMCISION
There were three main rationalizations in favor of male circumcision in traditional Nso': (1) Circumcision prepared the penis, putting it in a state of readiness for coitus and procreation, considered as the main purpose and *raison d'être* of marriage. (2) Circumcision tested the courage and endurance ability of a boy at the threshold of adulthood, during which these qualities would be indispensable and frequently needed. (3) Circumcision tames, moderates and tempers the sexual instinct, thereby helping a man to act responsibly. The traditional Nso' were very much aware that the pleasures of sex, like those of drink and food, are best enjoyed in moderation.

The above rationalizations for circumcision, thanks to historical evolution and development of Nso' culture, thanks to better critical awareness, would be quite unconvincing today, or, at any rate, considered insufficient to justify generalized routine circumcision. But, until fairly recent times, Nso' traditional society was one in which great endurance and courage were called for, especially from men, whose main occupations were warfare, internal security, hunting, house-building and long-distance perambulate trading, while women and children concentrated on home-keeping, farming and child care. Whenever any danger threatened in traditional Nso', the impulsive reaction was always quickly to remove the women and children to a safe place, and then all the men came out and faced frontally whatever danger it was. In fact, once a danger alarm was heard, a man grabbed his machete (in a scabbard) and spears, gave quick instructions to his wife (wives) and children, and dashed off in the direction of the alarm without any precise knowledge of what it might be all about. While discussing circumcision with a Nso' elder recently, he told me:

Yes, the uncircumcised penis is very sensitive. By simply rubbing it, an uncircumcised man can get the type of pleasure that you and I can only get by penetrating a woman. So, such a person may not feel the urge to get married and, if trouble comes and an alarm is sounded for men to come out, he may prefer to hide and rub his genitals; or else, he may become a real kibëv (he-goat), smelling any female from a distance and rushing madly after her.

It can be deduced from this remark that uncircumcised adults were not completely unknown in traditional Nso'. In fact, I am told that an adolescent boy who was very afraid of circumcision usually escaped to his maternal grand father (*taryiy*) where, by tradition, he would usually be treated like a prince and flattered and coaxed instead of ridiculed and bullied to submit himself for the ritual at the next opportunity.

INFANT CIRCUMCISION

So circumcision, within Nso' traditional society, was not without plausible rational justification. But, with colonisation and consequent introduction of Christianity and modern Western medicine, routine circumcision of male children at birth was instituted and has since become the uncritically accepted norm. Infant circumcision was at first, however, greatly resisted on the ground that it interfered with the growth and maturation of the penis, putting it in danger of remaining stunted. The Nso' particularly abhor any interference with the immature (Tangwa, 1996, p. 197).

There are no statistics available, but the rate of deaths from circumcision in traditional Nso' must have been appreciable, given that the operation was performed without any anaesthesia and that treatment of the wound was quite rudimentary. Such deaths were, however, considered as a misfortune (*lon*) and diviners were consulted to find out the cause of the cause of death. The belief in double causality is widespread in Africa and is usually dismissed by Westerners as superstition because it seeks explanations beyond what may be called the scientific causal chain. But, within the traditional

African world view, the fact that my child dies of malaria because s/he was bitten by a mosquito does not explain why, of all the children playing in the courtyard, the anopheles chose to bite but my own.

BACKGROUND AND RIDERS TO MY ARGUMENTS

The well mediatised Western campaign against female circumcision, led me to try to find out more about it from some medical doctors in Cameroon. Many practising doctors in Cameroon seemed to know little about it, but I discovered that the Cameroon Medical Women Association had investigated the issue in 1995 as part of a wider study focusing on 'traditional practices affecting the reproductive health of the women in Cameroon' (Wambua, 1995). They discovered that female circumcision was practiced amongst two marginal ethnic groups: the Ejagas on the western border with Nigeria and the Shua-Arabs of the extreme north on the border with Chad, but they were unable to determine the prevalence rate of the practice.

MALE AND FEMALE CIRCUMCISION

It is Doctor Joseph Jumbam of Saint Monica's Clinic, Nkar-Jakiri, who gave me some medical literature of studies carried out in Sudan, Nigeria and Somalia (Verzin, 1975; Asuen, 1977; Dirie and Lindmark, 1991) on the subject which greatly moved me and convinced me of the desirability of abolishing the practice. But when I started to think of the moral arguments involved, it became clear to me that, even though most types of female circumcision - which range from *circumcision* proper through **excision** through **infibulation** or so-called **Pharaonic circumcision** to **introcision** (the most drastic) - are evidently more severe than male circumcision, there is no strictly moral argument against female circumcision that would not equally apply to male circumcision. But I noticed that many people who readily admitted the necessity for the abolition of female circumcision or who even actively campaigned against it were completely silent over male circumcision or even sang its praises.

Early in 1997, I had informal discussions with some young Cameroonian ladies to whom I had lent some literature of the National Organization of Circumcision Information Centres (NOCIRC) sent to me by Marilyn Fayre Milos. I was quite surprised that, while they all readily condemned female circumcision, they all thought male circumcision not only good but necessary. They were unanimous that the uncircumcised penis looks 'funny' while the circumcised one looks 'elegant.' None of them was quite sure that she could bring herself to have sex with an uncircumcised man.

SEMINAR IN CAMEROON
In August 1997, the Cameroon branch of the Inter-African Committee on Harmful Traditional Practices Affecting the Health of Women and Children (IAC) organized a National Seminar on 'Female Genital Mutilation in Cameroon' which I attended as an observer. The information available on the seminar theme during this meeting did not seem to go beyond the 1995 findings of the Cameroon Medical Women Association although there were unsubstantiated claims that the practice was spreading. I raised the issue of male circumcision and tried to argue that, even though it might be considered less drastic than its female counterpart, it violated the same fundamental principle of non-respect of personal bodily integrity and that the two were therefore best tackled together, especially as it is the former that is widespread all over Cameroon. But, quite surprisingly, my arguments were rapidly brushed aside, as all other participants seemed to consider male circumcision quite unproblematic. Besides, as one of the organizers of the event pointed out to me, the terms of reference for which funding for the seminar was provided did not include male circumcision. A young lady even further told me during the coffee break that she considered my remarks typical of male patriarchal chauvinism, always seeking to divert attention from any serious problem affecting women to a non-problem affecting men. When I protested that I, in fact, considered myself a feminist, she only sneered.

TREATING EQUALS UNEQUALLY
This surprisingly ambivalent attitude towards male by contrast with female circumcision, a clear case of treating equals unequally, is very evident in the Western world where it may be attributed to both the 'Jewish factor' and cultural bias. Because of the importance of male circumcision in Judaism, and because of the importance and influence of the Jews in the Western world, Westerners are timorous of condemning male circumcision unequivocally. With female circumcision, it is much easier not only to condemn unequivocally but to propose the most draconian disincentives, because it can be viewed by 'enlightened Westerners' simply as one of those 'strange

and disturbing practices' of unenlightened non-Westerners (Lightfoot-Klein, 1997, p.131) who 'attack the genitals of their infants and young children with sharp knives' (DeMeo, 1997, p.1) because of their idiotic belief in hypocritical myths (Zwang, 1997, p. 67). It is significant to note that some states in the USA have lost no time at all in enacting legislation against female circumcision but have developed cold feet when urged, in the interest of fairness and consistency, to extend the legislation to cover male circumcision (Baer, 1997; Svoboda, 1997).

DISCLAIMERS AND QUALIFIERS

The following disclaimers, confessions, credos, specifications and qualifiers are necessary for a correct and balanced appreciation of this paper. This paper does not pretend to be a 'scientific study' or even a 'study' at all. These are just straightforward stories and narratives on the basis of which I attempt to argue for a point of view which I believe to be morally correct.

I was myself circumcised in infancy (contrary to the traditional practice of my people who never circumcised before adolescence) in accordance with a practice borrowed from Christianity and modern Western medicine. I am neither ashamed nor proud of my circumcised penis and, in fact, would consider any such pride or shame irrational. Singing the praises of the circumcised or the intact penis makes no sense to me. As far as I am aware, my circumcised penis performs its functions well, except those I may be blissfully ignorant of, since it must not be forgotten that altering the form may alter the function (Pollack 1997, p. 165). I wish I had been left intact to decide for myself, as a responsible adult, what reversible or irreversible interventions I wanted to perform or have performed on my body. But it is inconceivable that I would think of blaming my parents, let alone taking them to court to claim damages, for having arranged to circumcise me in infancy. I am convinced of their good intentions and sincerity of purpose, as I am of that of other parents (as I know them within my own culture), their epistemological and other limitations notwithstanding. I would not dream of seeking 'foreskin restoration', which, I suspect, would be, at least, as risky as circumcision itself. My arguments against circumcision are not based on the supposition that the circumcised penis is handicapped or that the intact penis is necessarily perfect. I believe that nature always has its reasons. It is not for nothing that nature made the shape of our toes different from that of our fingers. But that may not be a good reason to leave either our toe or our finger nails untrimmed.

In my youth I made a serious bid for the Catholic priesthood as a life-long vocation. While studying in the seminary, I discovered that African paganism and Western Christianity were very similar and quite compatible. No one else seemed to see it the way I did and I realized that I was a potential heretic in the Catholic Church, for which reason I abandoned the path to the priesthood. Reactions to my views on circumcision often remind me of that indelible experience of my youth.

GIST OF MY ARGUMENT
The gist of my argument is the following. Except for indisputably curative therapeutic reasons, infant circumcision ought to be abolished, because it clearly violates the fundamental ethical principles of bodily integrity, autonomy and self determination without the fully informed consent of the victim. While such consent, in some other cases, can be deemed to be reasonably given by the parent on behalf of the child, this should not apply in the case of circumcision because of its drastic nature, irreversibility and lack of urgency. Circumcision should have no place at all in paediatric medical practice.

FOR AND AGAINST CIRCUMCISION
The possible advantages of infant circumcision, as claimed by pro-circumcisionists, such as prevention of masturbation, insanity, hysteria, epilepsy, nymphomania, syphilis and other venereal diseases, penile and other cancers, alcoholism, bed-wetting, paralysis, malnutrition, nervous disorders, etc., even if granted, at best only balance its possible disadvantages - excruciating pain, sundry short and long-term complications, life-long trauma, removal of healthy harmless tissue with specifically clear and important functions, violation of intactness, etc., - as outlined by anti-circumcisionists. From the lay person's point of view, the 'scientific' evidence for and against circumcision is clearly inconclusive either way. While several recent scientific studies (Warren 1997, Hodges 1997, etc.) have debunked many of the claims of earlier equally scientific studies in favor of circumcision, new claims that cannot easily be brushed aside continue to be made. As recently as February 1998, for instance, Royce et al. (1998) of the Department of Epidemiology and the

Division of Infectious Diseases, University of North Carolina, Chapel Hill, and Family Health International, Research Triangle Park, make the following claim in a scientific report published in The New England Journal of Medicine (1997, 336: 1072-6) and diffused electronically on HEALTHNET NEWS (HIV/AIDS SUPPLEMENT), No. 29, February 1998, by Satelife, Boston MA, USA:

Male circumcision consistently shows a protective effect against HIV infection. This may be due to the abundance of Langerhans' cells in the foreskin or to a receptive environment for HIV in the sulcus between the foreskin and glans. The prevalence of HIV infection is 1.7 to 8.2 times as high in men with foreskins as in circumcised men, and the incidence of infection is 8 times as high. A greater proportion of the sex partners of uncircumcised men than of circumcised men are infected with HIV, which suggests that the presence of the foreskin may also increase infectiousness.

SCIENCE AND MORALITY

As Gajdusek (1992, p. 205) has so aptly remarked:

The scientist ... is expected by his peers and the public to report what nature reveals to him, and his observations and their reporting must in no way be influenced by issues of value, morality, personal or social gain, race, nationality, patriotism, religion or creed. His integrity as a scientist demands that. When the scientist communicates to the public or layman about the possible social impact of his work and of its moral, religious, economic or political consequences and implications, he is no longer only a scientist. His self-assurance and the certainty of all his assertions must yield, with humility, to the uncertainty of all matters human.

No scientist, however, can escape facing the moral issues involved in his/her work because the cardinal duty of reporting what nature reveals objectively, truthfully and disinterestedly is already in itself a serious moral issue.

A PERSONAL PREROGATIVE

As no rational being would knowingly and willingly act against his/her own better judgement, each adult person ought to freely decide for himself/herself which scientific claims and recommendations to follow. And, science apart, each adult human being, on the basis of the principle of autonomy and self-determination evoked above, ought to decide for himself/herself what personal risks (gambles) to take (make), provided these in no way adversely directly affect others. Supposing then that such an autonomous self-determining and well-informed adult freely decides, for whatever reason, to undergo circumcision, should s/he not have the right to have it done under the best medical conditions and facilities available and affordable? Consider for a moment, for example, the following putative example: A young adult American has as his/her single-minded life's ambition, tenancy of the White House. Both pro-circumcisionists and anti-circumcisionists agree on one point, namely, that circumcision appreciably reduces the sex urge. Might such a putative young ambitious adult, in the interest of a trouble-free future term of office as the Head of the most powerful nation on earth, and the sundry putative consequentialist benefits to his nation and humankind as a whole, not be justified in freely deciding to undergo circumcision? Is it not sometimes permissible or even morally heroic to do a lesser evil to achieve a greater good, especially if such is in the form of a personal sacrifice from altruistic motives?

RATIONAL PERSUASION

Besides, where circumcision has been truly a cultural or religious practice, its abolition cannot be sudden and automatic without severely disruptive consequences. Cultural and religious practices, because they have to do with community life and world-view, can only change gradually as other ideas, practices and habits supplant them. Rational persuasion, by contrast with threats and coercion, is an eminently ethical way of getting human beings to change their actions, behavior, customs or practices. Acting out of conviction is morally superior to acting out of constraint. A pupil who does his/her homework out of love for the subject is a better student than one who does so out of fear of the cane, even though they may both achieve the same good results. In like manner, I consider law to be inferior to morality. To act out of moral conviction is clearly superior to acting out of legal constraint, even if the result achieved, from the point of view of society as a whole, is the same. This is one reason that consequentialism and utilitarianism in general, which seek to evaluate the morality of actions by considering only their results or consequences, remain inadequate as ultimate moral theories, in spite of their general and widespread appeal.

It is particularly important that the fight against such practices as circumcision in Africa should not be propelled by Western threats of economic sanctions and withholding of development aid or even by incentives (Müller 1997, p.160). That would have the consequence of turning a moral crusade into the type of power game that the Western world has played against Africa since colonization. In this regard, Pollack (1997) and Goodman (1997) appear to me eminently on the right tracks towards contributing to the disappearance of circumcision, not only in Jewish circles- their immediate focus- but in the entire world. The fight against all types of circumcision must be premised on the fact that no human beings are perfect but that all are perfectible (Pollack 1997, p. 171). The fight against all types of circumcision cannot also be dissociated from the need to constantly rethink and redefine (name) the sacred, the need to constantly change our metaphors, the need to constantly check and recheck our actions and practices against life-affirming and life-protecting values, the need to constantly strive to 'humanize and rehumanize the world' (Goodman 1997, p.178). From this perspective, the fight against circumcision is not different from the fight against non-therapeutic abortion, the fight against torture and war, etc. If we think carefully, it becomes evident that we cannot (consistently) condemn circumcision while supporting non-therapeutic abortion or even such other practices as tatooing, facial marking, sterilization, plastic surgery, etc., when they are not dictated by any therapeutic necessity; nor vice versa. We cannot (consistently) support non-therapeutic abortion as 'every woman's fundamental right' and at the same time condemn neo-natal circumcision as a 'violation of every child's fundamental right,... a violation of the elemental maternal instinct to love and protect the child,... a violation of the entire life-giving feminine biology.' (Pollack, 1997,pp. 167-169). We cannot convincingly affirm life-giving and life-protecting instincts generally as the essence of holiness (Pollack, 1997, p.172) and at the same time support circumcision. For, if 'raising a knife to the genitals of a child is not holy' (Pollack, 1997, p.172), that is precisely because raising a knife to any other part of the child's body is equally unholy.

CONCLUSION

Circumcision for non-therapeutic reasons rightly deserves our condemnation; but we ought to equally condemn, with the same unequivocal breath, killing and maiming and torture, thanks to guns, bullets, land-mines, hand-grenades, tear-gas, water-canon and other weapons, manufactured and distributed for profit, with general connivance of silence. As we discuss any of these issues, we ought also to keep the others in view, for the time being at least with the corners of the eyes. If the linkage cannot be made, then, maybe, I am once more confusing Paganism and Christianity.

REFERENCES

Asuen, M I. (1977): 'Maternal Septicaemia and Death After Circumcision'. In *Tropical Doctor.* Pp.177-178.

Baer, Z., (1997): 'Circumcision: Are Baby Boys Entitled to the Same Protection as Baby Girls regarding Genital Mutilation?' In Denniston, G. C. and Milos, M. F. (eds.), *Sexual Mutilations: A Human Tragedy.* New York and London: Plenum Press. Pp.197-203.

DeMeo, J. (1997): 'The Geography of Male and Female Genital Mutilations'. In Denniston, G. C. and Milos, M. F. (eds.), *Sexual Mutilations: A Human Tragedy.* New York and London: Plenum Press. Pp.1-15.

Denniston, G. C. (1997): 'Circumcision: An Iatrogenic Epidemic'. In Denniston, G. C. and Milos, M. F. (eds.), Sexual Mutilations: A Human Tragedy. New York and London: Plenum Press. Pp. 103-109.

Derie, M. A., and Lindmark, G. (1991): 'A Hospital Study of the Complications of Female Circumcision'. In *Tropical Doctor.* Pp. 146-148.

Gajdusek, D. C. (1992): 'Scientific Responsibility'. In Fujiki, N. and Macer, D. (eds), *Human Genome Research and Society: Proceedings of the Second International Bioethics Seminar in Fukui, 20-21 March, 1992.* Eubios Ethics Institute. PAGE

Goodman, J. (1997): 'Challenging Circumcision: A Jewish Perspective'. In Denniston, G. C. and Milos, Marilyn M. F. (eds.), *Sexual Mutilations: A Human Tragedy.* New York and London: Plenum Press. Pp.175-178.

Laye, C. (1957): *The African Child.* Fontana Books. Place of publication

Lightfoot-Klein, H. (1997): 'Similarities in Attitudes and Misconceptions about male and Female Sexual Mutilations'. In Denniston, G. C. and Milos, M. F. (eds.), *Sexual Mutilations: A Human Tragedy.* New York and London, Plenum Press, Pp. 131-135.

Mzeka, P. (1996): *Rituals of Initiation in the Western Grassfields of Cameroon: The Nso' Case.* Reprinted from *Rites of Passage and Incorporation in the Western Grassfields of Cameroon. Vol. 1: From Birth to Adolescence.* Bamenda: Kaberry Research Centre, 1993.

Novak, M. (1970): *The Experience of Nothingness.* New York: Harper and Row.

Pollack, M. (1997): 'Redefining the Sacred'. In Denniston, G. C. and Milos, M. F. (eds.), *Sexual Mutilations: A Human Tragedy.* New York and London: Plenum Press, Pp. 163-173.

Svoboda, J. S. (1997): 'Routine Infant Male Circumcision: Examining the Human Rights and Constitutional Issues'. In Denniston, G. C. and Milos, M. F. (eds.), *Sexual Mutilations: A Human Tragedy.* New York and London: Plenum Press, Pp. 205-215.

Tangwa, G. B. (1996): 'Bioethics: An African Perspective'. In *BIOETHICS*, Vol. 10, No. 3, Pp. 183-200.

Turner, V. W. (1971): 'Symbolization and Patterning in the Circumcision Rites of Two bantu-Speaking Societies'. In Douglas, M. and Kaberry, P., (eds.), *Man in Africa. Garden City.* New York: Anchor Books, Doubleday & Co., Inc., Pp.229-244.

Verzin, J. A. (1975): 'Sequelae of Female Circumcision'. *Tropical Doctor.* Pp. 163-169.

Wambua, L. E. (1995): Traditional Practices Affecting the Reproductive Health of the Woman in Cameroon: An Executive Summary (Pamphlet).

Warren, J. P. (1997): 'NORM UK and the Medical Case against Circumcision: A British Perspective'. In Denniston, G. C. and Milos, M. F. (eds.), *Sexual Mutilations: A Human Tragedy.* New York and London: Plenum Press, Pp. 85-101.

Wiredu, K. (1990): 'Are there Cultural Universals?' In *Quest: Philosophical Discussions.* Vol. 4, No. 2.

Wiredu, K. (1995). *Conceptual Decolonization in African Philosophy*. Ibadan, Nigeria: Hope Publications.

Zwang, G. (1997). 'Functional and Erotic Consequences of Sexual Mutilations'. In Denniston, G. C. and Milos, M. F. (eds.), *Sexual Mutilations: A Human Tragedy*. New York and London, Plenum Press, pp. 67-76.

CHAPTER TEN

FEMINISM AND FEMININITY:
GENDER AND MOTHERHOOD IN AFRICA

[This chapter has been presented in several conferences and symposia over a number of years but has otherwise never been published before]

ABSTRACT

In this paper, it is my underlying claim that, although the Western feminist movement is largely responsible for the positive global shift in consciousness and attitudes towards women and the status of women, Africans do not need profession and/or practice of feminism to effect the emancipation and empowerment of African women. Western models and paradigms are globally very imposing and influential, especially in Africa – thanks to the cumulative impact of the various colonial legacies. But genuine non-alienating development in Africa calls for the use of all foreign and outside influences consciously as construction materials for an edifice whose foundation is purely African. Or else, there is the risk of ending up with an unstable structure whose foundation and center of gravity are not firmly in the ground but floating unstably somewhere above it. The oppression, suppression and marginalization of women in the Western world have been of a different texture, caliber and character from what can be considered to be similar or the same phenomena within other cultures. Furthermore, in attempt to liberate and emancipate themselves, Western women have acted and reacted in ways whose cumulative effect has been detrimental to femininity, family, heterosexuality and the role of the **woman as mother**, a role which is central in African culture and conceptions as defining and circumscribing the status of women. The emancipation and empowerment of African women can and should build upon traditional African foundations, where patriarchy could not validly be counter-posed to matriarchy and where the woman, as mother, wife, daughter or sister already had a revered and enviable status, even if this did not exactly match that of men in many respects. It is my contention that Africans could safely borrow the spirit but not the letter or agenda of Western feminism.

FEMINISM AND FEMININITY: GENDER AND MOTHERHOOD IN AFRICA[13]

People sometimes go as far as Sokoto in search of something they have in their sokoto.

(Nigerian adage)

INTRODUCTION

One of the more positive global developments of the out-going century has been a significant change in the over-all status of women, at least, at the conceptual level. There is no human culture in which women have not, at best, been marginalized and, at worst, suppressed and oppressed. No well-informed critical thinker would disagree with Rosaldo and Lamphere (1985, p. 3) when they state that

...most and probably all contemporary societies, whatever their kinship organization, or mode of subsistence, are characterized by some degree of male dominance ...none has observed a society in which women have publicly recognized power and authority surpassing that of men ....all contemporary societies are to some extent male-dominated, and although the degree and expression of female subordination vary greatly, sexual asymmetry is presently a universal fact of human social life[14].

In my view, this situation derives, ultimately, from a very simple fallacy or tendency in human thought according to which **might is right**, an attitude paradigmatic of Western culture but not completely absent in any other culture. The fallacy is responsible for the **abuse of power** in all its manifold forms and categories, of which physical strength and strike force are the most impressively palpable and convincing instances. In the case of the male-female

---

[13] I am very grateful to Dr. Ulrich Loelke of Hamburg University, Germany, and to Mrs. Alice Aghenebit-Mungwa of the Gender Information Valorization Facility – Africa (GIVaF), Cameroon, for very stimulating critical comments on the first draft of this paper.

[14] Quoted in Mineke Schipper (1991): *Source of all Evil: African Proverbs and Sayings on Women.* Chicago: Ivan R. Dee Inc. p. 5.

equation, the secret of the whole issue is the fact that, biologically, men are generally physically stronger, behaviorally more active and more aggressive than women. The fallacy is equally present in other power domains such as those of knowledge, economics, technology, information, etc., and the fight against it should be directed radically at the abuse of power in all its generic forms and manifestations.

However, today, at the threshold of the 21st century/3rd millennium, there is no part of the globe where it has not been accepted, at least in principle, that the status of women needs to be improved, that women need to be empowered to step out from behind the curtains and the shadows unto the centre-stage of socio-economic and political activism, for the benefit of both women and men. What has, perhaps, not yet been fully realized is that the struggle for the emancipation of women and for gender equality is co-extensive with the struggle against the abuse of power by both men and women generally in all domains, the struggle against economic exploitation and marginalization, racial and cultural discrimination, ideological indoctrination, religious, political and linguistic domination, etc.

FEMINISM AND DE-FEMINIZATION

The Western feminist movement has had a lot to do with the global shift in consciousness (or lack thereof), conceptions, perceptions and attitudes to women in our epoch, thanks largely to Western information, communication and mass education technologies. But, like many other things Western, feminism may be a double-edged sword, one of whose sides is liable to shred and stampede into the ground some valuable elements of other cultures, such as African culture, under the mistaken presumption or assumption that it holds the one and only key to understanding the problems of women and their solution, and to the empowerment of women, particularly. It seems to me that Western feminism, in spite of all that can be said in its favor, has succeeded largely at the expense of what I would call femininity (the sum total of positive feminine qualities/characteristics) and that African women need not sacrifice femininity on the altar of feminism to achieve empowerment.

Femininity, no doubt, connotes and implies certain attributes such as fragility, delicateness, vulnerability, prudence, subtlety, patience, etc. But these are only dispositional properties. That something is fragile, for instance, does not mean that it will break, much less that it should break. Dispositional properties need never be actualized, but the above dispositional properties of femininity, taken together, give their possessor all that can be captured in the indefinable concept of 'charm'. The de-feminization (if you would allow such a coinage) of women in the Western world has had a very direct adverse effect, as it seems to me, on the role of the woman as procreator and mother, a role liable to evoke the secret envy of men. For, although the father contributes a seed in the procreative process, this contribution is rather discreet and paternity remains largely a matter of blind faith and confidence in the mother, for both father and child. All men are, at best, only curious outsiders to the very fascinating mysteries surrounding human reproduction, mysteries which take place right inside the woman and by which she shares very directly in the unfathomable prerogatives of Nature/God. For that reason, the death of a woman, within Nso' culture, my natal culture, is mourned for four days whereas that of a man is mourned for only three days.

The Western feminist de-feminization of women is, however, quite understandable within its own context. Because of the male-oriented, male-centered, male-dominated character of Western societies, Western civilization has always valued attributes of masculinity higher than those of femininity; to the extent that the term 'man' in Western conceptions is deemed to be a generic term for all human beings, while 'woman' is a sub-set or specification of it. In this way, the male perspective is the perspective of all human beings on any given issue. Because of this, Western men have, understandably, never seen themselves as having a gender problem, which is seen as an exclusive affair of the women. In this light, it is not at all surprising that Western women should be fleeing from femininity and any talk of feminine attributes, which they perceive as man-made or invented by men in order to suppress and oppress women. Western feminists see feminine attributes as dispositions that have helped to keep women in bondage, imprisoned within man-made shackles, from which they, naturally, want to liberate themselves. For Western feminists, therefore, opposing femininity is part and parcel of a liberating impulse and imperative. But there is no reason why African women should borrow the historical baggage and psychological hang-ups of another culture and add to their own proper burdens.

NSO' CULTURE AS A PARADIGM

In this paper, I have limited my illustrative examples to a single African culture – that of the Nso' of the Bamenda Highlands of Cameroon – with which I have first hand familiarity. But I have used Nso' culture simply as a paradigm for mainly qualitative and prescriptive rather than quantitative and descriptive purposes. What is true of the Nso' here is very likely also true, *mutatis mutandis*, or, at least, would have appropriate parallels within other African sub-cultures, given their family resemblance. My aim is not a sociological or comparative empirical study of the position of women in African cultures. My generalizations are therefore not to be taken as inductive conclusions drawn from carefully collected and analyzed data but rather as prescriptive enunciations based on an argued position and demonstrable paradigms. The main thrust and direction of my argument and conclusions can easily accommodate or, at least, peacefully cohabit with particular counterfactuals. These qualifications are important because this paper can plausibly justifiably be accused of narrow sampling and selective bias. But a demonstration, for instance, that a medicinal plant for a particular disease exists somewhere within our own forest cannot be countered by saying that it is only a small and stunted one or that the rest of the plants in the forest are non-medicinal.

STATUS OF WOMEN IN TRADITIONAL AFRICA

The procreative/maternal role is central to the position and status of women in African cultures. And this role is, of course, only one arm of what Ali Mazrui (1990, p. 179) has termed a 'triple custody' whereby, according to African cosmogonic myths and metaphysical ideas, God made woman the custodian of fire, water and earth. Custody of fire entailed responsibility over energy, principally, in the context of traditional Africa, fuel wood; custody of water involved responsibility over a most important substance, water, symbol of both cleanliness and survival; and custody of earth implied a doctrine of dual fertility of soil and of womb. '*Woman ensures the survival of this generation by maintaining a central role in cultivation – and preserving the fertility of the soil. Woman ensures the arrival of the next generation in her role as mother – the fertility of the womb*' (*ibid*). These roles give the woman a central and pivotal position in all that concerns the being, well-being and prosperity of the community.

WOMEN OF THE BAMENDA GRASSFIELDS

In Cameroon, these traditional roles of the African woman are nowhere better illustrated than in the culture of the so-called grass-fields. In his Introduction to P. M. Kaberry's *Women of the Grassfields: A Study of the Economic Position of Women in Bamenda, British Cameroons* (1952), Darryll Forde (Director of the International African Institute, London), who, as early as 1945, had been charged by the colonial government of Nigeria to conduct a survey of the economic and social position of women in the Cameroons Province and in Bamenda in particular, makes the following very revealing and instructive statements:

The activities of the Bamenda woman are many-sided and cannot, without distortion, be abstracted from the context of tribal life and placed in a cultural vacuum. Their investigation entails, primarily, a study of the economic, kinship, religious and political institutions. The need for the same breath of approach holds equally for the presentation of the results of research. In other words, ones starting point is not the women but an analysis of a particular aspect of culture. ... this handling of data yields a series of statements on the role of women in particular segments of tribal life. Generalizations at this level are valid, significant, and of value to those concerned with the problem of raising the status of women and promoting their welfare. An attempt to go beyond this and, by a species of anthropological or moral arithmetic, to decide whether the position of women in general is high or low, or good or bad is, in my opinion, likely to prove profitless. ... Clearly, the replies to such questions are at best superficial, at worst distorted, and almost invariably contradictory. Let me give two examples which I found embedded in the Government files. A missionary who had spent many years in Bamenda asserted that " the women of this Division have achieved a remarkable degree of freedom and independence contrary to notions abroad". Another missionary of a different denomination but also of considerable experience of the country stated: "The status of women is alarmingly low. The main causes of this are considered to be the dowry system and polygamy-especially in its extreme form as practiced by local chiefs, village heads and prominent men of the tribe". As a further contribution to this diversity of opinion it is not inappropriate to cite those given me by some women of Nsaw. "Woman is an important thing, a thing of God, a thing of the earth. All people come forth from her." And this statement-also made by the wife of a polygamist: "A woman is a very God. Men are not at all. What are men?" Sometimes the answer to this rhetorical question was "worthless!" Again, in response to my inquiry why people mourned four days for a woman and only three for a man, the men offered the explanation: "A woman is one who bears the people (of the country). Women are very important. Women are like God, because they bear children." It should be noted, however, that to the people of Nsaw even God has his limitations and is not omnipotent.

Phyllis Kaberry herself, in her research among the grass-field women, came to the conclusions, *inter alia*, that, in spite of influences from the colonial impact, the women, as wives, mothers and daughters, effectively controlled the food economy – a domain of activity within which "they enjoy considerable independence and have well defined rights" (p. V111); that, in the whole Bamenda region, each married woman generally has her own dwelling (constructed for her by the husband), which she occupies with her unmarried children, whereas the husband, especially if he has no other wives, "may possess no special dwelling of his own, but when circumstances permit, he builds one in order to have space for his belongings and the entertainment of his guests. Adolescent sons may share this with him, but as soon as possible they take over a vacant hut and in any case they will, when they marry, construct one for the bride" (p. 6).

Kaberry highlights the importance of the maternal cult (*kitaryiy*) amongst the Nso', whereby the individual's health, well-being, welfare and success in life are believed to be connected with his/her mother's rather than father's lineage, with his/her matrikin and maternal ancestors (p. 13 &14). She particularly stresses the point that the casual European observer, "confronted with the spectacle of (Bamenda) women bending over their hoes through the day" (p. 27) are liable to draw the erroneous conclusion that there is inequitable division of labor and exploitation of the female sex and that the women are depressed and oppressed and have nothing "but care and woe" (p. 88) as their assigned lot in the scheme of things. In pre-colonial Nso' the woman was, in fact, what, in Western parlance, is called the bread winner of the home. Today's '*Massa, chop money don finish*' type of wife is a colonial legacy within Nso' culture. In traditional Nso', patriarchy and matriarchy were not in opposition, let alone conflict.

THE MATERNAL CULT IN NSO' CULTURE

While most African cultures could be roughly described as patriarchal, because of the preponderant role of men in public affairs and positions of a political nature, and because marriage, in general, is mainly patrilocal, though not always patrilineal, this is not because matriarchy is completely absent or unimportant. There are matriarchal societies in Africa and some cultural systems such as that of the Nso', for instance, can be rightly and accurately described as a patriarchal system, founded and grounded on matriarchy and the maternal cult. The Kingdom of Nso' was founded not by a man but by a woman, *Ngon Nso'*. And, although Kingship (*Vifon*, Fonship) is not open to females for mainly conceptual reasons, some women occupy the highly respected status and position of Queen-mothers (*ayaa*), a rank without any male equivalent, while the second highest politico-religious office, that of *Tawong* (ritual father of the Kingdom) is matched exactly by that of *Yewong* (ritual mother of the Kingdom). Queen-mothers command such awe and respect that even lineage heads (*ataala'* or *afai*) are wont to step off the road to the side to allow them respectful passage. And, on passing, she would often address the '*fai*' by his pre-installation name (*yir kiban*) – a privilege permitted only to clear superiors in rank – and he would respond: '*Yaa*' with his hand to his mouth, a sign of self-abasement and respect before a titled superior. Here, matriarchy and patriarchy are inextricably dovetailed.

The most fearsome masked juju in all of Nsoland is *kibaranko'*; if it happens to break loose from its restrainers, as it occasionally but rarely does, everyone, including its restrainers and attendants, flees for cover. But the greatest antidote of *kibaranko'*, even when it has gone berserk, is a pregnant woman. On sighting a pregnant woman, *kibaranko'* becomes completely disarmed and as shy as a bride in the presence of her in-laws and even a child can, at that point, lead it by the hand to its base in the secluded confines of *Nwerong la'*. Within this culture, the active and sometimes destructive male principle is moderated and controlled by the passive female principle. The apparent passivity of women does not necessarily imply privation of power. In a pot of soup, the salt, which is the most important ingredient, is completely invisible and never as flamboyant as the oil, which is quite dispensable.

AFRICAN WOMEN AND WESTERN INFLUENCES
If we take the grass-fields traditional setting as paradigmatic, African women had a status and general elbow-room scarcely matched by any other culture, at any time. If today the situation is somewhat different, it is largely owing to the contact with Western culture, the

colonial impact and, particularly, Western metaphysical and religious dogmas, ideas, customs, laws, practices, propaganda and assimilationist tactics. There is no doubt, for instance, that Christianity, in spite of its **one man one wife** dogmatic policy, and its occasional idolization and idealization of **woman as virgin** (complimented by men not born of any woman), has greatly contributed to the debasement and marginalization of women, especially colonial and post-colonial convert African women. In Christianity the unquestioning subordination and subjugation of women to men is raised to the status of an indispensable virtue and a divine command. The most respected and influential classical Western philosophers, from Plato through Aristotle, Saint Augustine, Saint Thomas Aquinas, Schopenhauer, etc., have also severally defined and described the woman as an 'imperfect man', 'an incidental being', 'a creature neither decisive nor constant', 'lacking certain qualities', 'afflicted with a natural defectiveness', etc. Any misogynist in search of theoretical support can find firm conceptual pillars and moorings in the Christian Holy Bible and the classical writings of the most famous of Western philosophers.

African women and men need to be very circumspect and imaginative and critically minded in their relationship with Western feminism because even a good idea uncritically accepted, line, hook and sinker, can turn out to be very deleterious and dangerous. There are many alibis for the imposition of Western ideas, systems, opinions, points of view, idiosyncrasies and preferences on others. These include presenting them as the most (or usually the only) rational, reasonable, ethical, scientific, sane, or workable ones; ultimately as the only ones ordained and commanded by the one and only true God, the Judaeo-Christian God. This situation is further exacerbated by the fact that, in virtually all domains of global human concern, the Western world not only has effective monopoly over the discourse, but ample resources for canvassing its points of view as well as possessing high persuasive skills which invariably evolve into lobbying when persuasion fails, change into bullying when lobbying proves unsuccessful, and finally transform into threat or use of naked force when all else fails (Tangwa, forthcoming). To bully

It is a remarkable fact that many of the really 'successful' feminists of the Western world have openly rejected heterosexual marriage and/or motherhood and their success, such as it is, is, in fact, thanks to their release from the constraints of wifehood and motherhood. Lesbians have been centrally active in the most important campaigns of the feminist movement (Smyth 1994, p. 22). And as Wilkinson and Kitzinger (1994, p. 75) rightly observe: 'feminists have produced searing critiques of heterosexuality, pointing in particular to the abuses associated with heterosexual sex'. I consider heterosexual marriage and motherhood to be of great value and importance to the human species as a whole. Equally important for the human species is the 'maternal instinct', which would cease to exist if motherhood, as we have known it, no longer exists.

Western feminism has tended to see empowerment of women mainly in terms of doing the same things that men do (and doing them better?) – from wearing trousers through standing to urinate to engaging in boxing and wrestling. Pushed to their limits, however, the demands of feminism would seem to be in principle un-fulfill-able, as long as the plain fact is recognized that a woman is not a man and a man is not a woman. As Wilkinson and Kitzinger complain: '... heterosexual sexual activity continues to reinscribe biological and social maleness and femaleness, constructing women as women and men as men, in order to ensure male dominance and female subordination. ...nothing changes: heterosexuality continues to mean what it has always meant under heteropatriarchy, and to serve the same sociopolitical function' (ibid, pp. 88-89). But I do not think that having gendered roles and occupations is necessarily a bad thing, as long as there is fairness and equity, and dogmatic exclusion and rigid monopoly are not given the status of unbreakable laws.

I agree completely with Ali Mazrui when he opines that economic differences are not the real explanation for the political subordination of women, since the same situation persists even where women are economically very active, such as in parts of Africa, but I thoroughly disagree with him when he goes on to suggest that the real problem lies with the military marginalization of women. According to Mazrui, 'what will one day change the balance between men and women is when the military machine becomes bisexual' (op. cit. p. 182). For the sake of both women and men, I would rather say that the military machine ought to be completely dismantled. Aggression and belligerence, which are predominantly male attributes, and which lead directly to war (the greatest avoidable man-made evil on earth) are attributes feminism should try to completely eradicate rather than equally share in.

There will always be the anomalous woman or man who feels more at ease doing some of the things that are normally associated with the opposite gender or who can assume any role effortlessly, in spite of its gender slot or associations. Biological facts are certainly the key to understanding gender differences but they do not inevitably establish a fixed and immutable destiny (Simone de Beauvoir, p.60). Sex is a biological fact but gender is both psychological and cultural. Gender could be contrary to what mere physiology might lead us to expect. For instance, '...although the external genitalia (penis, testes, scrotum) contribute to the sense of maleness, no one of them is essential for it, not even all of them together....gender role is determined by postnatal forces, regardless of the anatomy and physiology of the external genitalia. ...Psychosexual personality is therefore postnatal and learned' (Millett, op. cit. p. 30). What is necessary is to recognize the importance of all occupations and all roles, including the less spectacular ones. So far, the importance of the occupations and roles of wife, mother, child bearer, child rearer and attendant domestic chores have not been fully recognized in most cultures the world over. These are roles and occupations in which femininity and its attendant characteristics and attributes are indispensable. These are roles and occupations for which women are particularly though not exclusively suited, in the same way we can say that men are particularly but not exclusively suited for military combat. I see nothing wrong in socializing boys to be males and girls to be females. Such socialization merely cooperates with nature and natural propensities and, in both cases, should strive to eliminate negative and destructive tendencies. Nature will always produce some anomalies, but I see no reason why a woman would want to be a man or to acquire male attributes and characteristics, nor vice versa.

AFRICAN WOMEN WRITERS

Some of these considerations are no doubt what have, quite understandably, made some African women, no less desirous of emancipation and empowerment than their Western counterparts, to feel reluctant qualifying themselves as 'feminist'. For this reason, some African scholars have preferred discussing the emancipation and empowerment of African women under the concept of **womanism** as against that of **feminism** (Eboh, 1991). Some female African writers have also been quite reluctant to embrace Western feminism unequivocally. In an interview with Adeola James (1990, pp. 23-24), Ama Ata Aidoo of Ghana had the following to say on the woman question:

The fact is that women have been disregarded. In other societies, in other periods, at other historical points, people have managed to put women down and still operate a viable society. Quite obviously it is not the same for us now. Look at this vast continent! It is quite ridiculous, really, that people, especially educated African men, operate as though women were not around. That is part of the colonial inheritance, because it wasn't like that in our societies, at least not in most of them. Although, at every stage, women have not been given that headship position, our societies have not been totally oblivious of the presence and existence of women. I think it is part of that whole colonial rubbish that our men behave the way they do. I think it is about time that they woke up and we woke up and did something about it.

Zulu Sofola of Nigeria makes a very similar but less nuanced point in answer to the question as to whether African women writers can help to resolve some of the debates concerning our underdevelopment and the oppression of women by men.

I have always said that the only way that the African woman of today, with her European orientation which we call education, can be liberated, is to study the traditional system and the place of the woman as defined by it. There was no area of human endeavor in the traditional system where the woman did not have a role to play. She was very strong and very active. The only woman who has made history within the European structure was Mrs. Funmilayo Ransome-Kuti, but even she used the traditional system. Take the historical heroines – Moremi, Idhia, Queen Amina, even this Omu who we have just talked about; they achieved whatever they did through their traditional role or office. In the European system there is absolutely no place for the woman. In the traditional system the roles are clearly demarcated. For example, recently someone was telling me that a woman can never be an Oba (a traditional chief). So I replied that that was true, but the other side of the coin is that the Oba can never be the Iyalode (traditional head of women's affairs). The Obi can never be an Omu. The two positions exist, one cannot usurp the role of the other. The thing is that in the traditional system the human being was recognized as a human being (Adeala James, p. 150).

Oyeronke Oyewumi (1997) is analytically more refined. She demonstrates the inappropriate nature of Western feminist beliefs about sex and gender in other cultures, particularly African cultures.

According to her, the Western concentration on the **status** of women, whereby women are seen as powerless and disadvantaged, is quite misleading when applied to her natal culture – that of the Yoruba of Oyo, where women are not strictly defined in gender terms. In traditional Yoruba society, social status was determined, not by gender, but by social hierarchy that, in turn, was determined by social relations and the principle of seniority based on chronological age (p. 13). She further points out that the Yoruba language is a genderless language and human attributes are not in any sense gender specific (p. 174). In traditional Yoruba society, women were not differentiated from men or subordinated to them. The essential point here is true of many other traditional African societies, though expressed or manifested in a variety of different ways and forms.

THE CHARACTER AND TONE OF WESTERN FEMINISM
Western feminism is thoroughly defiant and confrontational and never has anything good to say about men or the Western patriarchal system. For that reason, a man in the Western world, who honestly believes in the emancipation and empowerment of women, is in a rather untenable position describing himself as a feminist without appearing to be either effeminate or hypocritical. In their Introduction to a very fascinating work, significantly entitled *STIRRING IT: Challenges for Feminism*, the editors boldly declare that the articles collected in the work 'not only deal with troublesome issues but also intend to make trouble' (Griffin et al. p. 1). They see stirring trouble as part of the legitimate 'business of feminists'. African women do not need to stir any trouble to effect their emancipation and empowerment because the abandoned foundations of such emancipation and empowerment are there in the traditional system and worldview, under the accumulated dust and debris of colonization and Christianization, waiting only to be salvaged and modernized. There is no need to create an artificial war of the sexes in Africa where none has ever existed as a precondition to the emancipation and empowerment of women. African women do not need to go to Sokoto in futile search of something that is in the pockets of their own sokotos.

One the most comprehensive, compelling and influential works of Western feminism is Kate Millett's *Sexual Politics* (1969) in which she tries to lay bare the biology, psychology, sociology and politics of the Western *war of the sexes*. She sees the family (in its

Western 'nuclear' conception, of course) as patriarchy's chief institution and weapon of oppression, separate from but organically connected with society and the state, which she considers even more patriarchal. For Millett, the patriarchal family has an essentially feudal character (complemented by the familial character of feudalism).

It is both a mirror of and a connection with the larger society; a patriarchal unit within a patriarchal whole. Mediating between the individual and the social structure, the family effects control and conformity where political and other authorities are insufficient. As the fundamental instrument and the foundation unit of patriarchal society the family and its roles are prototypical. Serving as an agent of the larger society, the family not only encourages its own members to adjust and conform, but acts as a unit in the government of the patriarchal state which rules its citizens through its family heads (ibid, p. 33).

This feminist attack on the family fortuitously accords very well with the rise, popularity and spread of male homosexuality, lesbianism, paedophilia, sodomy, contraception, abortion, and artificial methods of reproduction in the Western world. All of these, excepting only the last mentioned, are tabooed and considered as abominations in most African cultures; which is not to say that they are completely non-existent for, in that case, there would be no need for the taboo. These developments within Western culture do not augur well for the human family, human societies and the human species, as we have known them. For, no matter what can be said in defense or even in praise of some of these practices, (homosexuals, for instance, have been greatly and unjustly persecuted in some Western societies), the indisputable fact is that they are no aids to procreation and family life. They are, admittedly, what have helped most Western countries successfully to drastically reduce their population growth rate for the purpose of raising the general standard of living. But, while Africa urgently also needs to bring its population growth under more rational control for the same purpose, this can be done without borrowing or encouraging any of such practices.

Western technology has, furthermore, successfully transformed the mystery of human reproduction into a mechanical process of production controlled by corporate finance, free market forces and the calculative manipulations of the super-powerful nations of the world. Correspondingly, there has been a progressive loss of interest in the family, marriage, and natural heterosexuality. As Joseph Fletcher (1988) so eloquently states:

No longer is human reproduction centered in the genitalia or even dependent upon them. Even the gonads, testicles and ovaries are no longer necessary. ...The age-old saying omnes vivum ex ovo, all life comes from the egg, simply is not true any longer. ...Love making and baby making have been divorced. Sex is free from the contingencies and complications of reproduction, and sexual practice can now proceed on its own merits as an independent value in life (pp. 10 & 15).

AFRICAN CULTURAL FOUNDATIONS FOR WOMAN EMPOWERMENT

It is clear that what Western women are up against in their struggle for emancipation and empowerment may be quite different from what African women are up against. What Africans (men and women) and other colonized peoples have suffered in the hands of Western men (with the acquiescence if not connivance and complicity of Western women) should be an accurate indication of what Western women have probably suffered in the hands of their men. In this regard, it must not be forgotten that colonialism has, probably, destroyed the values, attributes and self-worth of African men more than those of African women. Empowering African women might also help to empower African men to find their lost identity and self-respect. In traditional Africa, women were not marginalized or suppressed and oppressed to the same extent or degree or in the same ways as women in the Western world or women in post colonial Africa. Furthermore, the maternal cult is present to varying degrees in all African cultures, which greatly value children and procreation.

All over Africa, a carrying child or a child in any sort of trouble is always heard crying out for the mother, even though it also has a father. And when a young person provokes another by calling his/her father names, the latter would normally promptly return the compliment by also calling the provocator's father names or, if s/he is a coward, s/he might smile and walk away in calm. But, if anybody says anything uncomplimentary about your mother, it is customary and expected of you to immediately launch into a furious fight, be you a coward or not, and no matter how you rate your strength and chances against your provocator. In an African setting, when the teacher or any adult is judging and correcting two kids who have been fighting, the one who struck the first blow is usually held culpable and severely reprimanded. But, if the reason for striking the first blow is 'He cursh my mami' or 'He cursh me, cursh my mami', then the other party would usually take the blame. What the musician, Prince Nico Mbarga, expresses in the lyrics of his song *Sweet Mother* is very convincing popular African philosophy: *You fit get another wife; you fit get another husband; but, you fit get another mother? Oh no!*

Western feminists have had to coin such interesting terms as 'shero' and 'herstory' in contraposition to 'hero' and 'history' respectively. But, in most African languages, these concepts have no gender bias. In Lamnso', for example, a s/hero is simply *ntav* or *wanntav* without any gender bias. Western culture is obsessed with gender differentiations to the extent that most Western languages assign gender to even inanimate objects. In most African languages, there is a single word or personal pronoun for the Western *le/la, he/she, er/sie*, etc. And, to take only the French language for example, it is very significant to note that, while it is highly obsessed about the correct gender form of all words, it has no feminine equivalent for nearly all the prestigious or lucrative professions and occupations: *scientifique, écrivain, professeur, médicin, ingenieur, architecte, magistrat, avocat, ministre, fonctionaire* etc. Evidently, *la culture, langue et civilization française* never envisaged women in any of these domains. Is such the culture from which African women and/or men hope to learn how to empower women?

Individuality, as distinguished from individualism, is also well recognized within African culture, to the extent that a married woman does not have to assume the husband's name, nor a child the father's name. The idea of Mrs. X, where X is a man's name, assumed by a woman because she is married to X, is a legacy of colonialism in Africa. At a feminism session during an international conference in Yaounde a few years ago, I listened with amazement to some participants counting with a celebratory air among the achievements of feminism the fact that it was now acceptable in Cameroon for a married woman, instead of completely dropping her maiden name and assuming that of her husband, to retain both by joining them with a hyphen. But what can any woman gain from conjoining the name of one male – her father's – to that of another male – her husband's?. The parallel practice whereby children assume the father's name produces some ludicrous results in the case of some gender-specific male names. Lamnso' names like Miss Wirba/Wirdzerem/Wirgo/Wirgham/Lukong, etc. are colonial legacies, which not only are ludicrous but radically subversive of Nso' metaphysics and worldview. African women can surely attain emancipation and empowerment without having to substitute a foreign worldview, culture, customs and practices for their own. Such substitution, unless first properly and thoroughly indigenized and domesticated, leads to alienation and inauthentic living.

CONCLUSION

My argument in this paper has not been either that African women do not need emancipation and empowerment, however these may be conceived, or that, in emancipating and empowering themselves, they should not borrow useful positive ideas from the experience of women in other cultures, particularly Western culture. In particular, it is not my claim, nor do I want to imply, that all is well with the situation of African women or that there was nothing wrong with the traditional status quo. My preoccupation has been rather with conceptual decolonization, procedural authenticity and self-reliance, which can only rest on traditional African values, culture, social norms, customs and practices. Acculturation and socialization, no doubt, sometimes lead to uncritical acceptance of unjust or untenable practices. But this is a problem for neither Africa alone nor women alone, but for both genders in all places and cultures. A decolonized, authentic and self-reliant approach requires a certain critical self-awareness that should reduce the danger of mistakenly importing rubbish or of needlessly begging, borrowing or buying from abroad what is freely and abundantly available at home. Such an approach should result in something of genuine and enduring value that can withstand and outlast current dominant Western fashions, obsessions and preoccupations and that can also influence other cultures positively.

REFERENCES

De Beauvoir, S. (1953): *The Second Sex*. Trans., H. M. Parshley. London: Jonathan Cape.

Eboh, M. P. (1991): 'The Woman Question: African and Western Perspectives'. Unpublished Paper presented at the World Conference of Philosophy, Nairobi, July 21-25, 1991.

Fletcher, J. (1988): *The Ethics of Genetic Control: Ending Reproductive Roulette*. Buffalo New York: Prometheus Books.

Griffin, G. et al. (eds.) (1994): *Stirring It: Challenges for Feminism*. London: Taylor and Francis.

Kaberry, P. M. (1952): *Women of the Grassfields: A Study of the Economic Position of Women in Bamenda, British Cameroons*. London: Her Majesty's Stationery Office.

Mazrui, A.. A. (1990): *Cultural Forces in World Politics*. London/Nairobi/Portsmouth (N. H.): James Curry/Heinemann Kenya/Heinemann.

Millett, K. (1969): *Sexual Politics*. London: Virago Press Limited.

Oyewumi, O. (1997): *The Invention of Woman: Making an African Sense of Western Gender Discourses*. Minneapolis: The University of Minnesota Press.

Schipper, M. (1991): *Source of all Evil: African Proverbs and Sayings on Women*. Chicago: Ivan r. Dee Inc.

Smyth, A. (1994): 'Paying Our Disrespects to the Bloody States We're In: Women, violence, Culture and the State'. In Griffin et al. (eds.), *Stirring It: Challenges for Feminism*. London: Taylor and Francis.

Wilkinson, S. and C. Kitzinger (1994): 'Dire Straights? Contemporary Rehabilitations of Heterosexuality'. In Griffin et al. (eds.), *Stirring It: Challenges for Feminism,* London: Taylor and Francis.

CHAPTER ELEVEN

BIOETHICS, CUSTOMS AND LAWS IN THE PRESENT SITUATION OF AFRICA[15]

[This chapter has been presented at several conferences but has otherwise never been published before]

INTRODUCTION

Positive law can be considered as a derivative or subset of ethics or morality[16] in the sense that any particular law or even any entire legal system cannot be adequately justified without reference to morality. The justification of morality itself is not, of course, unproblematic and has sometimes led philosophers into circular reasoning or an infinite regress, occasioning skepticism or despair. But such metaethical issues are outside the scope of my present concerns. An immoral law has no justification, no matter the power and authority of its maker. Not only is it justifiable (even heroic) to defy or challenge such a putative immoral law, it is also even quite plausible to consider it as being no law at all.

There are, of course, positive laws which are *amoral* or morally neutral, such as the law to stop when the traffic light is red and to move only when it turns green. There is nothing of moral relevance if such a law were reversed in such a way that we stop with the green and move with the red light. With a morally relevant law this sort of arbitrary reversal of the rule would not be possible. Again, if you "beat" the traffic lights, you have committed a legal infraction but,

---

[15] I thankfully acknowledge the support of the *Alexander von Humboldt-Stiftung* for this work. This paper was first read during an international conference on *Bioethics and Biolaw* organised by the *Danish Centre for Ethics and Law in Nature and Society*, under the patronage of UNESCO, at Copenhagen, Denmark, 29 May - 1 June 1996.

[16] In this paper I make no distinction between morality and ethics although I consider such a distinction possible, ethics being, in my view, a derivative or subset of morality which stands to the latter like logic to thinking or theology to religion. I do not want to get bogged down here with semantic issues. For an illuminating distinction between morality and ethics, see Joseph Omoregbe (1979): *Ethics: A Systematic and Historical Study*. London: Global Educational Services. P. 3.

otherwise, done nothing morally wrong. Thus, while not every law is of moral relevance, an immoral law is absolutely unjustifiable. Furthermore, while a human society without laws is perfectly conceivable, one without morality would seem to be both a conceptual and practical impossibility. Laws ought to be moral but morality needs no legalization. In fact, I would go as far as saying that the morally more mature and advanced any society is the less is the need for positive laws in such a society. Laws, as well as customs and traditions, are essentially aids and short-cuts or rules of thumb to morality. They help us so that we do not have to reason and deliberate in an ad hoc manner each time we are faced with a moral question, problem or dilemma, because reasoning and deliberation are not easy activities. Ethical rules and even moral principles can also be considered in like manner as aids, in ascending order of generality, to morally right actions.

The moral system of any society grows out of its metaphysical beliefs, historical experiences, physical/psychic environment, its customs and practices, etc. No human community is perfect and, hence, no system of morality is perfect. But this imperfection does not imply the impossibility of progressive improvement or that of judging and/or comparing one moral system with another or any moral system with itself at different historical epochs. Our ability to deliberate upon, discuss, judge and/or argue about a moral law or ethical injunction or an entire moral or legal system clearly shows that human rationality is quite indispensable to morality although this does not imply that rationality itself is fixed and static. It is quite plausible to think that morality improves along with improvements in human rationality. The moral system within which "an eye for an eye and a tooth for a tooth" was a notable and acceptable ethical injunction or law was not without rational justification in so far as it did not prescribe two eyes for one or a row of teeth for one tooth. But there is no doubt that the succeeding moral system which replaced this *lex talionis* with "love your neighbour as yourself" (*lex Christiani?*) was more humane and more rational than its predecessor. Today, clear progress in both rationality and morality is being made by attempts to extend our understanding of the "neighbour" we are to love beyond the familiar fellow next-door, beyond ethnic group, tribe, race and nation, even beyond the human species (thanks to Peter Singer and his followers), to embrace all sentient creatures.

Bioethics may be, more generally, considered as dealing with ethical questions relating to the entire biological world and, more narrowly, as focusing on ethical questions within the context of modern medicine and biomedical research, as far as human beings are concerned. The narrower conception has developed side by side with remarkably rapid progress, within the western world, in medicine, biological research and technology. However, such progress has also been matched by the development of new and defiant forms of deadly diseases and global environmental threats to the entire biological world. Because of the rapidity and sheer novelty of some of the problems and dilemmas that arise within this context, it is not surprising that existing laws and traditional legal systems are often caught unawares and unprepared in this situation. Hence *biolaw* as a response, within a legal framework, to this new and urgent situation.

Because the biological and medical developments that bring about novel ethical problems and dilemmas have generally occurred in the western world, biolaw has, not surprisingly, at least, attempted to develop side by side with them in the western world. African cultures, world-views, philosophies, customs and practices which, traditionally, were, in the general sense, evidently bioethically very sensitive, have not yet awoken to the new bioethical challenges for fairly obvious reasons, the more palpable of which are that, since the phenomenon of western colonization, Africa has remained a mere consumer of western ideas and products and that the developments which make bioethics and biolaw contemporary imperatives have not yet been truly felt in Africa. And yet, African traditional systems might indicate or suggest better or more viable ways of tackling bioethical challenges in a world whose cultural frontiers are rapidly shrinking. If African traditional systems are looked upon, as they should, as viable **alternatives** rather than, as they usually are, as mere **approximations** to western "ideals", they stand to contribute great values to an emergent world culture made possible and inescapable by western advances in technology and communication.

THE STATE OF AFFAIRS IN PRE-COLONIAL AFRICA

In traditional Africa, before the western colonial intervention, metaphysical conceptions, ethics, customs and laws, ably assisted by taboos, formed a single continuous piece in such a way that it was not usual to separate or make distinctions between them. In my paper "Bioethics; An African Perspective"[17] I have attempted, *inter alia*, to describe an African world-view, that of the Nso', by contrast with the dominant western world-view. This world-view, which I have described as **eco-bio-communitarian**, is pluralistic and flexible in all its details by contrast with the western counterpart which, at best, is notoriously dualistic and, at worst, tyrannically monistic. Western dualism and binary thinking is clearly reflected in such famous dichotomies as the One/the Many, Realism/Idealism, Universal/Particular, Matter/Mind, Rationalism/Empiricism, Egoism/Altruism, Autocracy/Democracy, Capitalism/Socialism, Individualism/Communalism, Anthropocentricism/Biocentricism, Patriarchy/Matriarchy etc. Western monistic thinking finds eloquent expression in such dogmas as **monotheism** and **monogamy** which are normally presented, not as western beliefs and practices, but rather as universally and eternally valid and desirable ideals of rationality to which all human beings ought to aspire. In a recent article in NEWSWEEK[18] the author, who attempts to advance rational arguments for gay marriages as a promotion of monogamy, refers to polygamy as a "horror". Westerners would not be struck by the cultural myopia, if not blindness, and prejudicial arrogance underlying such a remark. Most Africans would consider the idea of same-sex-marriage a veritable horror. The difference, however, is that Africans have no global platform from which to express and canvass such opinion. In my paper "Democracy and Development in Africa: Putting the Horse Before the Cart"[19] I have attempted, *inter alia*, to compare the general ramifications of western binary thinking and monistic impulses with those of African pluralism and diversity.

The traditional world-view within which I was born (that of the Nso' of the grassy highlands of Bamenda in Kamerun) recognizes the differences between plants, animals, human beings and super-human spirits, but it does not imbue human beings with any mandate

---

[17] Forthcoming in *Bioethics*, Volume 10 Number 3 1996.
[18] June 3 1996, p. 20.
[19] I was to have read this paper at the AFRICA CONFERENCE on "Philosophy, Politics and Development in Africa: Assessing the Twentieth Century and Looking Forward", Binghamton University, USA, 7-9 June, 1996. But, failing to obtain a visitor's visa into the USA, I sent it to the organisers of the conference to be read on my behalf.

or special privilege, God-given or otherwise, to subdue, dominate and exploit the rest of creation, as is the case with the western outlook. The Nso' attitude towards nature and the rest of creation is that of respectful co-existence, appeasement and containment. Hence the frequent offering of sacrifices to God, to the divine spirits, both benevolent and malevolent, to the departed ancestors and to the sundry invisible and inscrutable forces of nature. The Nso' Year (*Ya' Nso*) always begins with fertility sacrifices (for plants, animals and humans) officiated by the King (*Fon*) of Nso', the chief High Priest of the Nso' Kingdom, assisted by three ritual concelebrants: *Yewong* (ritual mother of the Kingdom), *Tawong* (ritual father of the Kingdom), and *Fai Ndzëndzëv* (chief great councillor of the Kingdom). Through the course of the year, numerous other sacrifices are offered all over Nso' by the custodians of arable land (*ataangvën*), lineage heads (*ataala*) and *agaashiv* who normally combine the functions of medical doctor, priest, psychiatrist, psychologist and exorcist.

Among the Nso' even something like the treatment of illness is basically a matter of appeasement and containment. The aim is never eradication, elimination, wiping out, or extermination, but rather to coax and plead with the illness to leave its innocent victim alone. Otherwise, there is no attitude of insistence that illness, on its own, should not exist.

I still remember how as a young lad I once accompanied my mother during the harvesting season to pick up the leftovers of corn (*siyri rümi*) after the men had carried home the harvest the previous day. We picked a basket-full of corn and my mother said we should return home. I protested that there was still a lot of corn in the *rümi* (the spot at the centre of the farm where the corn is usually heaped by the women harvesting for the men to carry home). But my mother told me: "We never pick everything; we must also leave something for the birds and the rodents." "The same birds and rats who had been stealing the crop even before it was ripe?" I asked, incredulously. "Yes", replied my mother, "they too need to eat."

It is also significant to note that, before dishing out food (usually *kiban*, that is, *corn foofoo*) for the household, a traditional Nso' woman would always first break a small piece of it and throw it carelessly through the door or window. That is supposed to be for any hungry spirits so that they would allow the household to eat their own meal in peace.

There is a remarkable story that Reverend Father Patrick Isichei once told me about a community reconciliation ceremony (following some religious conflicts) which he witnessed in the *Ishan* area of mid-western Nigeria. As I can recall his account, the Catholic priest first prayed for the community, calling on God in his bountiful mercy, and in the name of his crucified son, Jesus Christ, the Prince of Peace, to draw the warring factions closer to himself and grant them peace. The Imam prayed next, calling *Allah* by his 99 names and asking for his peace on the warring community because Islam itself is synonymous with peace. Then it was the turn of an elderly African Traditional priest. He removed a kolanut from his pocket, peeled it and broke it into small pieces with his thumb nail. Then he threw a piece to one side and called on God to restore peace to his community. He threw a second piece on the other side and called on the ancestors not to abandon their children. Finally, he threw another piece behind him and called on the devil:

> *Devil, here is your own share of the kola. We do not understand the reasons for the quarrel between you and God because we were not there when it started. If the two of you do not like peace between yourselves, try and understand that we want it amongst ourselves. We beg you not to stand on our way to peace and harmony.*

The *Ise*! (Amen) from the whole crowd was deafening, clearly showing that all those present firmly identified with the prayer.

These stories may sound trivial but the attitude underlying them is very important. It is an attitude of giving even the devil his due. If foreign influence, interference and insistence on formalism and legalism were to allow, this attitude would surely lead the South African *Truth and Reconciliation Commission*, under Bishop Desmond Tutu, to find reconciliation, healing and lasting peace for all South Africans, in spite of their horrendous past. In *Lamnso'* we call it *kimann-* a public ritual of repentance and confession followed by forgiving and forgetting, to the extent of tabooing even any further mention of the affair. That way, the process of healing is greatly assisted. The purpose is not to condone atrocities. There is no doubt that this attitude is what has helped Africans and black people the world over to survive and recover from the horrors of slavery, colonization and racism and to forgive their inflictors. The process is therapeutic, salutary and salvific for both victim and victimizer.

It is indeed amazing that, more than 50 years after the Second World War, war criminals are still being pursued and discovered and "brought to justice" in the western world. What justice can be exacted from a single individual who exterminated hundreds of completely innocent human beings? I believe that this insistence on formalistic and legalistic justice only helps to keep alive the very deep wounds of the horrendous atrocities of the past. This western attitude, which I also suspect to be related to a deep-seated belief that "might is right" is what, in my opinion, has stood on the way of reconciliation, peace and security in such places as Bosnia, Chechnya, Northern Ireland, the Middle East, etc.

Among the Nso' the earth (*nsaiy*) is considered to be a very potent force, to the extent that it is sometimes identified with God. The earth is where all living things originated and where they all end up again. The Nso' do not joke with the earth at all because it is the scourge of all secret offenders and the final arbiter of all human affairs. The law of the earth (*nsër nsaiy*) is the ultimate and strongest and brooks no breaking because the consequences of its violation are severe, metaphysical and unavoidable, even if it is violated accidentally or unknowingly. The Nso' prefer to leave certain offences and offenders to the judgement and sanction of the earth. In traditional Nso' land disputes were relatively unknown even though no land certificates or anything of the sort were issued for the plots on which people farmed or built. It was considered extremely dangerous to covet land to which you were not entitled because the spirits and forces of the particular piece of earth would not recognize you and you could not therefore sacrifice or pour libations on it. Using such a coveted piece of land spelled a lot of danger not only for the individual concerned but also for his/her whole extended family whose other members could not stand idly by while their collective well-being and existence were wantonly put into jeopardy.

Land tenure customs in Nso' were very loose but, nevertheless, extremely effective. In principle, the Fon was the titular owner of all land in Nso'. But in practice, all land for both farming and building was controlled by the arable land custodians (*ataangvën*) and lineage heads (*ataala*), respectively. Anybody who wanted land for farming or building could obtain it free of any charge by following the right procedure and approaching the right person.

Another important area where the coalescence of ethics, custom, and law is most evident in traditional African cultures and ways of life is **marriage**. Marriage is, no doubt, a cultural universal; but the **form** of marriage is clearly culture-bound. In western cultures, marriage is generally understood as a union between two individuals (a man and a woman?) to the **total exclusion of all others**, till death do them part, while in African cultures it is understood not in terms of **exclusion** but rather **inclusion** of many others. In Nso', as elsewhere in Africa, polygamy, monogamy and celibacy were practiced within a plastically flexible framework that balanced both freedom and rigidity.[20] There was no need to take a vow or sign a contract that you would remain in your chosen present marital status for the rest of your life.

Both in its polygamous and monogamous forms, there are three main roads to a valid marriage in Nso': orthodox marriage (*vidjin*), unorthodox marriage (*kincem* or *nceemi*) and wife inheritance (*villem*).[21] In orthodox marriage, the first or only daughter of a properly married woman is always given out in marriage by her maternal grandfather (*taryiy*) and all the other daughters by their paternal lineage head (*tarla'* or *fai*). A father arranges the marriage of the first wife of each of his sons as well as the marriage of each of all the first or only daughters of all his daughters. A peculiarity of Nso' marriage, *vis-a vis* marriage amongst other African peoples, is that there is no bride price/wealth or dowry. Ceremonial gifts are, however, exchanged at prescribed times with the proviso that the total value of marriage gifts received on behalf of a girl should not exceed the total value of the gifts that were given on behalf of her

---

[20] For a good description of such plasticity and flexibility in the marriage laws and customs of another African peoples, see Ulrike Wanitzek, "Between Continuity and Change: The Marriage Laws of the Bulsa of Northern Ghana", in COMMISSION ON FOLK LAW AND PLURALISM, Papers of the Commission's X[th] International Congress, University of Ghana, Legon, Ghana, 1995.

[21] For a more detailed description of these and other aspects of Nso' traditional marriage, see L. S. Fonka, "Nso Marriage Customs" and W. Banboye, "Methods of Wife Acquisition", a Nso' History Society Production, Nooremac Press 6, Abiodun Street Mushin, Lagos, Nigeria (undated).

mother. A lot of time and attention are, however, devoted to checking the familial background of the prospective spouses to make sure that there is no blood relationship between them and that there exist no old and unsettled grudges, quarrel or feud between the two family lineages. In Nso' marriage between blood relations up to the fifth degree is taboo. The prospect of marriage is also an occasion and an incentive for the ritual settlement of persisting differences, grudges and problems between families, lineages and communities.

In unorthodox marriage, a man takes unto himself a wife and a woman a husband without the above formalities of orthodox marriage. They simply move-in together. Children begotten of such marriages belong, by Nso' custom and tradition, to the parents of the woman, the man being considered in this situation as a mere "he-goat" (*kibev*), having no claim whatsoever to his own biological offspring. Furthermore, female children begotten through this form of marriage cannot marry in orthodox fashion but must also marry, if at all, in unorthodox fashion. This was a great disincentive and constraining factor for this form of marriage. However, many of such marriages, if they proved stable, happy and successful, could, retrospectively, be orthodoxified by performing all the requirements of orthodox marriage, some, obviously, merely symbolically.

In *villem* (wife inheritance) a man inherits the widow of his father, son or brother as his wife. When the occasion arises, it is usually the lineage head, in consultation with the women and elders of the compound, who decides who is to inherit whom. This form of marriage served as a sort of insurance and security for married women and had the consequence of reinforcing polygamy.

COLONIAL AND POST COLONIAL AFRICA

One of the long-lasting effects of the colonization of Africa is the rupture of the harmony between metaphysical conceptions, ethics, customs and laws. Western colonialists imposed their laws all over Africa in total disregard of the world-view, moral sensibilities, customs, traditions, practices and exiting laws of the colonized peoples. Of course, they also attempted at the same time, through formal education and religious proselytisation to inculcate a different world-view and an ethic that could better support the laws and projects they had imposed. The result has been total chaos all over Africa, compounded by the fact that, while the cultural and ideological background of all the colonialists was the same, the positive laws they imposed on Africans differed from place to place according to the linguistic and national peculiarities and pragmatic pre-occupations of the particular group of colonialists.

In Kamerun, which was first colonized by the Germans (1884) and subsequently carelessly unevenly divided as a First World War booty (1919) between Great Britain and France, and then partially re-united (1961) after a UN-conducted plebiscite, so-called Napoleonic (Civil) Law cohabits uneasily with English (Common) Law and, at some levels, Customary Law in some parts and even Islamic Law (Sharia) in others.[22] The net result has been the uprooting of law from its metaphysical, ethical and cultural foundations and its increasing dependence on the whims and caprices of executive power-wielders. Since the imposition of a "unitary state" in 1972, there has been a concerted effort to subvert and destroy the authority of traditional rulers and institutions in those parts of Kamerun where these are still important factors to be reckoned with. Such efforts culminated in Decree No. 77/245 of July 1977, which sought to turn all "traditional chiefs" into puppets of the civil administration by making them subordinate and accountable, even in traditional matters, to the "Senior Divisional Officer" in their area of jurisdiction.[23]

Cyprian Fisiy has recounted one telling effect of this state of affairs on land tenure laws, customs and practices.[24] According to this account, the Fon (King) of Ndu had, in 1974 or thereabouts, given an extensive piece of land to the Baptist Mission in Kamerun, in the true spirit of traditional laws and customs, for the construction of a primary school, a secondary school and a teacher training college. Afterwards, the Baptist Mission changed its mind about establishing

---

[22] For some legal problems that arise out of this situation, see Charles Fombad (1991): "The Scope for Uniform National Laws in Cameroon". In *Journal of Modern African Studies*, 29 and "The Motor Accident Compensation Ordinance, 1989, of Cameroon" in *Journal of African Law*, Vol. 39, No. 2, (1995).

[23] For a detailed historical background as well as attempted apologetic rationalisation and justification of this decree, see Paul N. Nkwi (1979): "Cameroon Grassfield Chiefs and Modern Politics". In *PAIDUMA*, Wiesbaden, Franz Steiner Verlag, 25.

[24] Cyprian F. Fisiy (1995): "Chieftaincy in the Modern State: An Institution at the Crossroads of Democratic Change" in *PAIDUMA*, Stuttgart, Franz Steiner Verlag, 41 (1995), p. 54f.

two of the proposed institutions in Ndu and transferred them elsewhere. Later, the Fon of Ndu requested to have back part of the land that was now lying fallow for redistribution to others for other purposes. But the Baptist Mission refused to release the land, claiming ownership over it, which it supported by procuring a "land certificate" from the civil administration. The case dragged on for nearly a decade and was finally decided in favor of the Baptist Mission on the ground that, since it had already "registered" the land and had a certificate for it, the Fon could no longer claim "the property". According to the pathetic account, the ageing Fon openly wept and lamented: "Who now owns land in Ndu and in the Wiya clan?" To which the agents of the civil administration gave the prompt response: "The Fon owns Ndu and the Wiya (clan), but the government owns the land and everyone who lives on it, including the Fon!"

Another sphere where traditional African culture had something of great value that has been almost completely destroyed is in the area of **eugenics** and **gender** considerations. While it may be generally true that women have been oppressed, suppressed and discriminated against in all cultures at all times, it is much less true in African culture than in western culture. Obsession with gender is very much a part and parcel of western culture and way of life as can be clearly seen from its cosmological myths, languages and practices. God in western conception is clearly a man and angels, though supposedly genderless, all bear male names. This is not the case in African cultures where female gods and divinities were not unknown and where there is always a single personal pronoun for s/he. Western feminists, in their struggle against patriarchy, have had to coin such terms as "shero", "herstory", "gynocentric", etc., coinages which would be absolutely unnecessary in African languages. In *Lamnso'* a s/hero is *ntav* or *wanntav* (son/daughter of a *ntav*). The Nso' system may indeed be generally described as patriarchal, but this is patriarchy based and founded on matriarchy, as the ubiquitous *kitaryiy* institution can confirm.

Furthermore, African culture and practice, in spite of its communitarian outlook and preferences, recognizes individuality (not to be confused with individualism) more than western culture and practice. Among the Nso', for instance, every child is supposed to come into the world with his/her own name which the family and community only try to guess (*shu' yir wan*). The widespread practice today whereby children bear their fathers' names and married women their husbands' names is one of the legacies of colonialism in Africa. The practice produces some very funny results in Nso' where, nowadays, a girl might be "Miss Wirngo" or a married woman "Mrs. Wirnkar"-manifest contradictions in terms and meaning in *Lamnso'*. (In *Lamnso'*, *Wirngo* and *Wirnkar* are the male equivalents of the female names *Wingo* and *Winkar*.).

As regards the upbringing of children, dominant western practice would seem to be characterized by exaggerated love, pampering, indulgence and monopoly of the child by the nuclear-parents, up to the age of about 8, followed by, apparently, total abdication and abandonment of all parental responsibility and authority. Under western influence, it is not uncommon today in Africa to find young children who can boldly tell their parents to don't be silly on an appropriate occasion. A few generations ago, it was inconceivable for an African child to tell the parents or even any older person to don't be silly. And a clear positive value in cultural borrowing has been achieved here, provided African parents don't abdicate the responsibility to bring up their children properly, including not refraining from thrashing them, if and when that is the only way left to get common-sense into their heads. But again, as all Africans know, *you don't beat a child with a big stick.*

CONCLUSION

Bioethics and Biolaw are two contemporary areas of concern where the advantages and even inevitability of a global outlook are easily evident. In this globalization, the western world is in a powerfully dominant position, thanks to its scientific/technological sophistication, especially in the area of communications and its economic prosperity. Non-western cultures, especially African cultures, can, however, make valuable contributions in this process, if they are given a fair chance and not brushed aside with characteristic impatience and prejudicial arrogance. People in the western world invariably tend to look on their own metaphysical conceptions, customs, practices and even idiosyncrasies as universal imperatives of human morality or rationality. Such outlook may not be unique to the western world, but it is particularly dangerous because the western world possesses the means and ability to spread and enforce its own outlook. No one can deny that the western world has found the right technology. It is doubtful that it has found the right way of **using** that technology or the best way of living as human beings. A society where you could live for nearly two years, as I have now done in Germany, without knowing your next-door neighbor, without making any personal friends (or even enemies) in your neighborhood, as is the case in many western societies, can surely not claim to have discovered the right formula for living as human beings. As a stranger in western society, you are so lonely and love-sick you could go outside and romance a stone, but nobody would care. Western culture and civilization has a big mouth and a loud voice which it knows how to use very well and small ears which are scarcely ever used. There will surely be a turning point for the better in the history of humanity and the world when western culture starts really listening to other cultures.

CHAPTER TWELVE

BIOETHICS AND INTERNATIONAL BIO-MEDICAL RESEARCH FROM THE POINT OF VIEW AND PERSPECTIVE OF AFRICAN CULTURE AND PHILOSOPHY

[A French version of this paper, under the title „Bioethique et recherche biomedicale internationale sous l'angle de la philosophie et de la culture africaine", is published in **Ethique de la recherche et des soins dans les pays en developpement,** *edited by Francois and Emmanuel Hirsch, Paris: Vuibert, 2005, pp. 55-75]*

ABSTRACT

One area in which the process of globalization has had an indisputable impact, for good or ill, is in the domain of bio-medicine. In recent years, there has been a great increase in the number of biomedical research studies in the developing world, especially in Africa. These researches are mostly carried out by researchers from the industrialized developed world, in some cases in collaboration with local scientists and researchers, but in all cases with funding from the developed industrialized world. Biomedical research in the industrialized world is very well regulated by a framework of institutional, national and international guidelines, backed in most cases by appropriate legislation and a generally enabling environment. In the developing world, however, such regulatory frameworks are largely inexistent. This situation raises grave dangers of unethical research, particularly so as extant international ethical guidelines, formulated in and by the industrialized developed world, from the viewpoint and perspective of the industrialized developed world, do not, in many respects, adequately address the interests and concerns of the developing world in an appropriate manner. In this paper, I attempt highlighting some of these problems and concerns, from the background, viewpoint and perspective of African culture and philosophies, as well as hinting in passing on how, in my view, the situation might begin to be redressed or at least addressed and ameliorated.

BIOETHICS AND INTERNATIONAL BIO-MEDICAL RESEARCH FROM THE POINT OF VIEW AND PERSPECTIVE OF AFRICAN CULTURE AND PHILOSOPHY

INTRODUCTION

The 10/90 gap (Global Forum 2001-2002) in all its ramifications is responsible for the present situation of the African continent vis-à-vis biomedical research. That situation is one in which the continent seems to present the most compelling conditions for externally driven biomedical research. These conditions include poverty, a heavy external debt burden, ignorance or lack of awareness, a heavy burden of all types of diseases and medical conditions, absence of any constraining regulatory frameworks, hopelessness and desperation, etc. So attractive have these conditions been to market/profit driven medical research that there seems to be a second scramble for Africa, comparable to that which led to the partition of Africa by European imperial powers in 1884. Other motivating factors for developed world's medical research in the developing world are altruistic philanthropy and the desire to bridge the 10/90 gap as a matter of global rational self-interest.

For these reasons, there has been an explosion in the number of biomedical research studies in Africa as well as in the other so-called developing regions of the world, particularly in the past two decades. Some of such studies have been accompanied by abuses and malpractices reminiscent of those which, within the Western industrialized world had led to the elaboration of regulatory texts for biomedical research, such as the *Nuremberg Code, Declaration of Helsinki, International Ethical Guidelines for Biomedical Research Involving Human Subjects, The Belmont Report,* etc. To cite only a single illustrative case, a clinical trial designed to text a drug called trovafloxacine alias Trovan, was recklessly carried out by a big Western pharmaceutical company on children in Northern Nigeria in 2001, during a meningitis epidemic, and resulted in the death of eleven of the children while a further two hundred became blind, deaf or lame. The case is briefly reported, amongst other similar cases, in *Developing World Bioethics* (Vol.1, No. 2, November 2002, p. 92) and discussed in some detail in a forthcoming article by Ruth Macklin (Macklin 2003).

The whole situation has been greatly compounded by the HIV/AIDS pandemic which can be said to have chosen sub-Saharan

Africa as its headquarters and against which a cure and/or vaccine is an extremely urgent necessity. Africa is thus caught in a situation where avoiding bio-medical research (whether internally or externally driven) is not an option in addressing the possible dangers of abuse or malpractices that may accompany such research. The only option is squarely to face the dangers posed by intensified bio-medical researches and to take the necessary steps and precautions to avoid them. Such necessary steps and precautions include appropriate ethical guidelines for medical research involving human beings, a system of regulatory frameworks at different levels, from the continental, through the national to the community or institutional, the building of capacity and expertise in the domain, and the creation of awareness and an enabling environment. There are formidable hurdles on the way to these objectives, but they are very worthy and compelling objectives.

CONCEPTUAL FRAMEWORKS

The different conceptual frameworks and ideational background against which modern bio-medical problems arise in the Western world by contrast with the African world need to be briefly sketched here for initial appreciation. Without a proper appreciation of such differences, it is difficult to conceptualize different scenarios or have a feel for what may be the appropriate solutions to the problems arising in each domain.

The most remarkable attributes of the African continent are diversity, variety and coexistence of a plurality of greatly differing entities and elements. Africa is one of the richest and most variegated continents on earth: ecologically, geographically, climatically, biologically, historically, culturally, linguistically, natural-resources-wise etc. Almost any ecological niche, taken at random, anywhere in Africa, presents remarkable and fascinating diversities and variety. In Cameroon, my own country, for instance, with a land surface of only circa 475,000 km^2, and a population of less than 15 million inhabitants, there are over 240 distinct linguistic/cultural groupings which have all experienced German and English or German and French colonization and influences, and the ecology, the flora and fauna are truly representative of what is available elsewhere in Africa.

Africa also happens to be the continent that, in the past as in the present, has been and continues to be exploited by the powerful predatory peoples and nations of the world.

African ethical, religious and metaphysical ideas have, over the ages, been influenced, shaped and coloured by this background of diverse data.

The pre-colonial traditional African metaphysical outlook can be described as eco-bio-communitarian (Tangwa, 1996, p.192.), implying recognition and acceptance of interdependence and peaceful coexistence between earth, plants, animals and humans, by contrast with the Western outlook which could be described as anthropocentric and individualistic. Within the African outlook, human beings tend to be more humble and more cautious, more mistrustful and unsure of human knowledge and capabilities, more conciliatory and respectful of other people, plants, animals, inanimate things, as well as sundry invisible/intangible forces, more timorous of wantonly tampering with nature, in short, more disposed towards an attitude of live and let live.

African philosophies, belief and ethical systems, both in their latent and forensic forms, provide the supporting structures for this world-view. Within the world-view, the distinction between plants, animals and inanimate things, between the sacred and the profane, matter and spirit, the communal and the individual, is a slim and plastically flexible one. For instance, the belief that humans can transform into animals, plants such as trees, or forces such as the wind, is very rampant within this belief system and has very significant implications for the way and manner nature as a whole and in its various specific manifestations is approached.

In like manner, metaphysical conceptions, ethics, customs, laws and taboos form a single unbroken piece within this African outlook, to the extent that, in the traditional past, it was not usual to separate them, at least from the point of view of the ordinary members of the community. They all covered what ought or ought not to be done, although sanctions for violations varied accordingly, the severest, perhaps, attaching to taboos, whose violation, whether voluntary or not, attracted consequences not only on the violator as an individual but, sometimes, also on his/her extended family or entire community and called for ritual atonement, cleansing and purification.

RESPECTFUL COEXISTENCE

The particular African traditional world-view within which I was born (that of the Nso' of the grassy highlands of Bamenda in

Cameroon) recognizes the differences between plants, animals, human beings and sub/super-human spirits, but it does not imbue human beings with any mandate or special privilege, God-given or otherwise, to subdue, dominate and exploit the rest of creation, as is the case with the Western outlook. The Nso' traditional attitude towards nature and the rest of creation is that of respectful coexistence, appeasement and containment. Hence the frequent offering of sacrifices to God, to the divine spirits, both benevolent and malevolent, to the departed ancestors and to the sundry invisible and inscrutable forces of nature. The Nso' Year (*Ya' Nso'*) always begins with fertility sacrifices (for plants, animals and humans) officiated by the King (*Fon*) of Nso', the chief High Priest of the Nso' Kingdom, assisted by three ritual concelebrants: *Yewong* (ritual mother of the Kingdom), *Tawong* (ritual father of the Kingdom), and *Fai Ndzëndzëv* (chief great councilor of the Kingdom). During the course of each year, numerous other sacrifices are offered by the custodians of arable land (*ataangvën*), lineage heads (*ataala'*), and healers (*agaashiv*), who normally combine the functions of medical doctor, priest, psychiatrist, psychologist and exorcist.

Among the Nso' treatment of illness is basically a matter of appeasement and containment. The aim is never eradication, elimination, wiping out, or extermination, but rather to coax and plead with the illness to leave its innocent victim alone, such innocence being a strong presumption which in some cases stands in need of ritual verification through divination. Otherwise, there is no attitude of insistence that illness, on its own, should not exist. If illness did not exist, how then would people die? In Nso' conception, death, as an inevitable end, is not considered a necessarily bad thing, especially when it is a timely and relatively painless one, neither premature nor overdue (Tangwa, 1996, p. 195). For the rest, the diagnosis and treatment of illness within the traditional culture were completely free of charge, except in so far as the patient or relatives might be asked to provide certain ingredients required for preparing his/her medicine or except in so far as s/he or they would usually voluntarily come back, as a matter of custom, after successful treatment, to 'thank' the healer with appropriate gifts.

The general point about this African background is that it is consistent with only a cautious and piecemeal use of technology or tampering with nature in general. It can accommodate biomedical research on human beings only if such research scrupulously respects the dignity and well-being of the research subjects and is geared towards solving palpable medical problems that have defied solution.

WESTERN CONCEPTIONS AND WORLDVIEW

The Western world has, perhaps, undergone more rapid and fundamental changes in its metaphysical conceptions, ideologies, moral thought, way of life and practices, than any other human culture. Such rapid changes may make it more difficult to identify or accurately describe the Western worldview, but that is not to say that such a worldview does not exist. Generally, it can quite plausibly be said that 'liberalism and possessive individualism' correctly describes the outlook and ideology that rules and moves the Western world.

Among the catalysts of change in the Western world, the *Industrial Revolution* of the 18th and 19th centuries has pride of place. The *Industrial Revolution* drew its impetus from the slogan that 'knowledge is power', convertible into commercial value; from the idea that all knowledge is unqualifiedly good; from the belief that nature is, in principle at least, completely knowable and controllable; and from perception of the universe as something which ought to be explored, subdued, dominated and exploited. These ideas and the attitudes they created had their origins and foundation in Judaeo-Christianity, but they led, paradoxically, to secularization, desacralization and profanation of everything in the universe - the very antithesis of the Judaeo-Christian ethic. The Industrial Revolution and the technologies resulting from it greatly assisted Western imperial nations in their voyages of exploration, discovery, subjugation, colonization, domination and exploitation of other peoples. The cumulative effect of these very important achievements was to infuse in Western culture, a culture underpinned among others by the ideology of rational self-interest, a spirit of omnivorous discovery, an automatic impulse towards unifying, patenting monopolizing and commercializing such discoveries and a penchant for spreading and promoting its ideas, vision, convictions and practices under the guise of universal imperatives of rationality and morality which ought to be binding on all and sundry.

This situation is further exacerbated by the fact that, in virtually all domains of global human concern, the Western world not only has effective monopoly over communications technologies and thus over the global discourse, but also possesses high persuasive skills, which invariably evolve into lobbying when persuasion fails, change into bullying when lobbying proves unsuccessful, and finally transform into threat or use of naked force when all else fails. Today, Western culture is, indisputably, the dominant culture of the world. That is only a descriptive fact to which neither blame nor praise is attributable; but Western culture is also a domineering culture, a situation that need not be and that remains amenable to prescriptive change. Particularly in the domain of science and technology, Western culture is the acknowledged master at whose feet other cultures sit as pupils and apprentices. But this does not mean that others also have to learn and accept all other things from Western culture. Science and technology in themselves have nothing to do with how, for example, people conceive of or worship God, how they marry or bring up children, how they organize their social system and, most importantly, science and technology in themselves have nothing to do with the USES to which people may or ought/ought not to put science and technology.

THE WESTERN WORLD AND OTHER WORLDS

The spirit of omnivorous discovery which the Industrial Revolution engendered and made possible in the European ancestors of what today is called the Western world guided them to all parts of the globe where they discovered peoples and cultures so different from theirs that they somehow felt reluctant to qualify them as 'human'. From then on, Europeanization (Westernization) of other peoples and cultures naturally appeared to them as humanization and civilization. It is in this way that both altruistic and egoistic motives became mixed and confounded in the relationship between the technologically very advanced Western world, peoples and culture and other (technologically less advanced) worlds, peoples and cultures. Since the Industrial Revolution, technology has been propelled to great heights by Western commerce and the profit motive, by war and the will to dominate, by pure epistemological and scientific curiosity, and to a lesser extent by the altruistic urge to improve human well-being. This situation has led to the marriage between medicine and commerce or 'the market', a fact fraught with both positive and negative aspects and consequences.

The Western world today is one which admits no taboos or forbidden areas to study/research, thereby allowing the spirit of individual acquisition and the profit motive, greatly aided by highly efficient technologies, to rule the world and even the entire solar system. If we venture to discuss these issues from an African perspective, it is only because, as Kwasi Wiredu has very aptly pointed out (Wiredu, 1994, pp. 34, 37, 38), technique is different from knowledge while knowledge itself does not necessarily imply wisdom and all, therefore, stand in need of moral appraisal, which requires no other credential beyond a good will.

The main underlying problem with the Western background and outlook is its epistemological over-confidence, bordering on arrogance and often resulting in recklessness. The constant perennial overturning and correction of past 'knowledge' by new 'knowledge' has not necessarily led to tentativeness by Western scientists and technologists in their claims or to more circumspection and caution in their interventions in nature. On the other hand, and by contrast, the main underlying value of the African world-view and outlook is its epistemological humility and respectful caution, as befits fallible beings, even if it also has the strong tendency toward conservatism and lack of progress.

BIOMEDICINE AND ETHICAL GUIDELINES
The issues in Western biomedical research and practice that have recently been most widely discussed include, randomly, the following: experimentations on human subjects, cruelty to animals (including using them in experimentation and killing and eating them), animal liberation/rights, informed consent, euthanasia/assisted suicide, artificial prolongation of life, brain death, transplantation of human organs, sale of human organs, cadaveric harvest/banking/sale of human organs/eggs/sperm, surrogate motherhood, artificial insemination, in vitro fertilization, abortion, acquisition/use of genetic information, creation/release of genetically modified organisms, animal cloning, human cloning, etc). Apart from those of these issues which concern non-humans (plants and animals), the rest could be roughly categorized into three groups: those relating to the beginnings of human life/birth, those connected with the duration of human life/living, and those concerning the end of human life/death. Biomedicine and biomedical research today intervene at all these levels, with human life as the irreducible permanent stake.

Apart from the complexities brought into these issues by overt and covert economic/commercial motives, the main underlying ethical issues involved here are connected with the mechanization of life and death or of their natural processes. This mechanization threatens to turn human reproduction, as we have known it, into mere production, and living and dying into similar processes depending largely on mere material calculations involving, inter alia, the reading of graphs, charts, balance sheets, insurance policies, patents, costs, and the turning on and off of machines. Western culture can be described as a technophile culture although that is not to say that there are no technophobes in the Western world. Without being either technophilic or technophobic, African culture could show Western culture the way back to those natural human values that Western culture has apparently sacrificed to the gods of industrialization, technology and commerce.

Beginning with the Nuremberg Code, all international ethical guidelines, notable among which are: *Declaration of Helsinki* (WMA 1964, 1975, 1983, 1989, 1996, 2000), *International Ethical Guidelines for Biomedical Research Involving Human Subjects* (CIOMS 1993, 2002), *Convention for the Protection of Human Rights and Dignity of the Human Being with Regard to the Application of Biology and Medicine: Convention on Human Rights and Biomedicine* (Council of Europe 1997), *Note for Guidance on Good Clinical Practice* (ICH 1996), have as their main concern the protection of the life, health, dignity and human rights of the human subject of medical practice or research.

But biomedical research in the developing world, especially in Africa, faces many challenges, dilemmas and difficulties including trying to conform and comply with some of the ethical imperatives of such research as laid down in the regulations and stipulations of the international regulatory texts, such as the need to avoid harm, exploitation and undue inducement or to respect autonomy and confidentiality.

In a community in which people do not keep any important secrets from their kith and kin, under pain of being suspected of having been initiated into witchcraft, for example, the idea of *confidentiality*, so much emphasized in Western biomedical ethics and in the international regulatory texts, is practically inoperative in its Western form and boils down essentially to *trust* and *confidence* ensuring *reliability*. In such a community, the sense of protection that confidentiality is supposed to give might be subverted by such metaphysical beliefs as witchcraft or lack of familiarity and/or trust in researcher or research. In such a community or similar other, the Western biomedical ethical dogma of, say, **autonomy**, again well reflected in the regulatory texts, may boil down essentially to **respect for other human beings** as moral equals.

Furthermore, in a situation of generalized poverty and hopelessness, combined with a high burden of disease, as we today have in some parts of sub-Saharan Africa, there is no way to prevent victims and potential victims of a deadly epidemic such as HIV/AIDS from being unduly induced by any type of research participation proposal that holds out the slightest possibility of any type of treatment. This situation is responsible, on the one hand, for what has been termed the 'therapeutic fallacy' by which research and other investigative procedures may deceptively be presented as if they were therapeutic interventions and, on the other, for what we may call the 'therapeutic illusion' by which research subjects labor under the incorrigible mistaken belief that they are undergoing medical treatment as opposed to study. In such a situation, the prohibition of **inducement**, due or undue, has no practical applicability and, if strictly observed, would amount to a prohibition of all medical research, which may not be ethical or to the best interest of those in need, all things properly taken into consideration.

This, however, does not mean that autonomy, confidentiality, the need to avoid undue inducement, etc. are not universal ethical imperatives; it only means that such imperatives, in their practical concrete *in situ* application, need to be sensitive to context, perspective and other constraining particular data. In such a situation, developed world researchers may and need to be pragmatic, provided such pragmatism does not involve doing what they themselves honestly believe or consider morally wrong. Ethics and morality are first and foremost about what I/we ought/ought not to do and only secondarily about what others ought/ought not to do. One does not change one's moral convictions and do what one considers morally wrong on the grounds that those with whom one is dealing think or do differently. If I consider adultery morally wrong, I would not become an adulterer simply by sojourning amongst people for whom free exchange of spouses is no big deal.

AFRICA AND HIV/AIDS VACCINE RESEARCH

HIV/AIDS vaccine research has become crucial for the world at large and absolutely critical for Africa at this point. The best resources for such research, however, are located, like most other strategic resources, outside of the African continent, specifically in the leading Western countries; hence the importance of ethics and general moral sensibility in addressing the problem. And, since the stakes here are extremely high in moral, human, material and economic terms, politics and politicking have become an inevitable part and parcel of the whole process. In this situation, international guidelines, rules and injunctions regulating biomedical research, which had hitherto apparently operated with little or no problem, have suddenly proved inadequate as guides for HIV/AIDS vaccine research, thereby occasioning a spate of discussions, revisions, and updates, amidst controversies. For example, both the *Declaration of Helsinki* and the CIOMS *International Guidelines for Biomedical Research Involving Human Beings* have been recently revised and updated in 2000 and 2002, respectively. The new *Revised Declaration of Helsinki,* which was unanimously adopted by the 52nd World Medical Association (WMA) General Assembly in Edinburgh, Scotland, in October 2000, clearly indicates the extent to which unambiguous clarity in the formulation of these injunctions and guidelines is felt to be necessary. Of the 32 paragraphs of the old *Declaration,* only 3 remained completely unchanged while the rest have been changed or modified and 8 completely new ones have been added. As Dr. Human (2000, p.1) says in his Introduction to the new *Declaration,*

It is the nearest that the Declaration has ever been to the Nuremberg Code and the UN Covenant on Civil and Political Rights in putting research subjects' interests first, and ahead of the interests of the researcher and of society in the outcome of the research. If it commands the wide international acceptance of preceding versions – and it should do, given the solid ethical basis of the changes – it will alter the climate in which most medical research is carried out, and may lead to some radical reassessment of what research is worth doing at all.

It is not surprising that some of the most controversial debates surrounding these revisions are connected with issues relating to developed world vaccine research in the developing world. In fact, one of the original motivations for starting on a revision of the Declaration of Helsinki came from the Americans who wanted a relaxation of some of the requirements of the Declaration so that they could carry out vaccine research in the developing world without violating the provisions of the Declaration. Ironically these demands have not been met by the new revised Declaration, which successfully maintained the solid moral basis on which it was first formulated, and the Americans have launched a strong lobby for its re-revision! The lobby against certain articles of the revised Declaration of Helsinki (notably 29 and 30) is thought to be a lobby on behalf of some multi-national pharmaceutical companies and the drug industry. I have had the opportunity to listen to some of their carefully elaborated arguments and it seems clear to me that what they are doing is basically presenting what are typical economic and pragmatic arguments in the language of morals. It can only be hoped that both Helsinki and CIOMS, these two important, time-honoured and internationally respected documents for ethical conduct in the domain of heath care and health research would maintain their solid moral foundations and remain resilient to attempts to confuse, merge or conflate ethics with economics and pragmatism.

HIV/AIDS AND CONSPIRACY THEORIES/EVIDENCE

As soon as the AIDS epidemic was first clinically reportedly noticed among homosexual men in North America and Europe in the early 1980s, accusations and counter-accusations about its source and origin started flying around the world (Sabatier 1988). In a residual cold war mentality, the Russians in 1986 claimed that AIDS originated as a result of American germ warfare. Some Western scientists speedily located the origin of the HIV, the virus that causes AIDS, in central Africa, without tracing a clear and convincing causal chain. No one paid adequate attention to the very pertinent remark of Kenneth Kaunda, former President of Zambia, in October 1987, to the effect that it was much more important to worry about where the disease was going to than about where it came from. Subsequently, blame for the spread of the epidemic shifted back and forth from male homosexuals through drug addicts, ethnic minorities ("with their queer sexual habits"), Haitians, Hispanics, Blacks, foreign students, seamen, prostitutes, Africans, Europeans, Western scientists, etc. The conspiracy theories, claiming that the virus was deliberately created as a weapon against certain groups of people, were, however, better worked out and presented, with at least *prima facie* plausibility (Cantwell 1988 & 1993; Horowitz 1996; Luinggvist 1992).

In a placing on the internet website UNIF-EVANGELISM@LISTSERV.AOL.COM, in July 2000, one of the theories about the origin of HIV/AIDS is said to have been elaborated by Edward Hooper in 1999 in a book entitled *The River*. In this book, Hooper, a former reporter for the BBC in Africa, allegedly claims that an oral polio vaccine used decades ago in Africa provided the transmission route for the HIV or a related virus from African chimpanzees to humans. The vaccine, the theory contends, was made with chimpanzee cells infected with simian immunodeficiency virus (SIV), which experts generally believe gave rise to HIV. It is claimed that the virus accidentally entered humans as a result of a massive experiment on a new oral polio vaccine conducted between 1957 and 1960 in the former Belgian colonies of Africa – Congo, Rwanda, Burundi. In the said experiment, American and Belgian scientists are said to have vaccinated nearly one million Africans, at a time that research/health ethics was not yet anybody's particular preoccupation. The polio vaccine in question – CHAT – is suspected of having been contaminated by SIV from chimps whose kidneys were allegedly used as tissue cultures in the manufacturing process. It is further pointed out that the vaccinations were conducted in precisely the towns and villages of central Africa where the earliest cases of HIV/AIDS are said to have developed. Hooper's own general conclusion is said to be that, even with the best of intentions, the rush to do genetic engineering experimentation is very dangerous as it is liable to succeed in curing one illness while at the same time inadvertently sowing the seeds of greater health calamities. The Hooper theory draws strength against dismissive incredulity from a similar theory reported on the website EUROELSAV@egroups.com in September 2000, said to be elaborated in a book entitled *Darkness in El Dorado* in which James Neel, a geneticist, who headed a study project among the Yanomami people of Venezuela in the mid 1960s, is accused, even if without any clear causal evidence, of using a virulent measles vaccine that sparked off an epidemic that in turn killed thousands of people.

Other people have gone much farther in claiming that the AIDS virus was deliberately synthesized, mass cultured and spread through vaccination programmes as a weapon of chemical and biological warfare and/or population control of Western multi-national fascists, bent on global domination and massive reduction or extermination of undesirable populations. The higher prevalence of AIDS amongst such groups as American Blacks, Latinos, Native Americans, Male Homosexuals, Asians and Africans etc. is pointed to as corroborative evidence to this theory. In the particular case of Africa, it has further been pointed out that a study of the map of small pox infection for 1967 shows a remarkable similarity to today's map of HIV infection. Those parts of Africa that had the highest rates of small pox and where vaccinations against small pox were concentrated are said to be the same that have the highest HIV infection rates. One of the most recent reechoing of these conspiracy theories is to be found on a May 7th 2003 placing on the website www.camnetwork@yahoogroups.com.

It is hard to say at this point whether these conspiracy theories should be entertained at all or rather dismissed off-hand as being too farfetched and fantastic and liable to distract attention from the real fight against the HIV/AIDS pandemic. One reason for taking them with a healthy pinch of salt but not dismissing them out of hand is that they tie in well with certain obsessions with population control, even if a direct causal link has not yet been established. It may, however, be too hasty or too uncritical to completely discount the possibility of such a causal link at this stage. Remarks like the one reportedly made by the renowned demographer and former United Nations Population Fund (UNFPA) representative, Jan Fransen, at a briefing for the European Parliament in Brussels in October 1999, to the effect that AIDS is *helping* to do the work of population control in Africa (EWTN News Brief, 18 October 1999), ought to make any critical thinker to stop and try to think between the lines of such a 'hilarious' joke. The hilarious joke came in the course of outlining three possible ways by which population growth in Africa could be limited, viz: fertility control, migrating people out of Africa, and increasing mortality. The joke was continued in the remark that no politician would campaign on the platform of the last mentioned possible way, which, of course, is not the same thing as saying that no politician could contemplate taking that possible way.

Conspiracy theories may, therefore, be pushed aside to the corner, for want of sufficient evidence, but they should be kept within retrievable reach, in case any further new evidence might warrant their reappraisal. After all Nazism and its propagated and attempted genocidal project were no fiction but recent historical facts and the contemporary world is not without the likes of Adolf Hitler, genuine power-hungry extreme racists and bona fide lunatics, some of who occupy positions of influence, power and responsibility.

The analytic arguments and practical reasoning at the basis of the obsession with world population control is nowhere more succinctly or better expressed than in a paper by John B. Hall (1996) published in the Journal *Population and Environment*. Hall starts with the relatively uncontroversial observation that the success of human beings in harnessing and exploiting the natural environment has led to such an unprecedented increase in the number of human beings depending for support on the natural environment that the limits of natural life-support systems appear to have been exceeded. It is therefore necessary, he continues, that the numbers of peoples consuming the world's resources be brought into some sort of balance with the available resources. In his view, the present world population is already far greater than can be sustained for much longer, even at 'present levels of misery'. Some of the consequences he draws from this state of affairs are: famine, war, ethnic strife, disease, all of which, in his view, lead to an impoverishment of the spirit and a fatal tendency towards group insularity, coupled with indifference to the suffering or death of others.

Hall's descriptive analysis is Malthusian in its thrust, if not inspiration, and contains little or nothing that is really new or controversial. It is his prescription of a solution that throws a shadow of doubt over his entire project. For him, the solution lies not only in ending further growth of the Earth's population but in actually and substantially decreasing the number of people the Earth is required to support. This necessary decrease in population size, in his view, would be most unlikely to come about voluntarily, as no nation or cultural group would like to believe that it is dwindling in size and no country would want to feel that it is losing population and that its own people might soon be replaced by fecund foreigners who are clearly all too ready to move into its relatively "empty spaces". Furthermore, Hall continues, while it is obvious that a decrease in population size is necessary to prevent impending disaster, many cultures still value high fertility levels and many unaffordable generations may be needed to change such attitude. Again, it is assumed that individual control over reproductive decisions is an inviolable and basic human right. But, reasons Hall, if the exercise of such a right is leading to universal disaster, why should it not be reconsidered and modified? His recommendation, therefore, is that, using modern genetic technology, an infectious contraceptive, which could spread as spontaneously through the population as the common cold, should be manufactured and used. This, as he tries to show, can easily be done, using the human immune system and certain existing viruses and antigens. Darwinian processes would then, presumably, take over to determine the select few who may successfully resist such infection. According to him free choice in reproductive decisions is one freedom we cannot afford, if we are to preserve any of the other freedoms that we need and cherish.

THE KISSINGER REPORT

Hall's recommendation, which by his own admission: "raises profound moral and ethical questions", would, perhaps, not be considered so alarming if seen as the isolated opinion of an individual. But, if connected with other concerns or read against the backdrop of some other controversial policies, it could assume added importance or significance. Some thoroughly horrified reader of Hall's paper, who claims to have had a long telephone interview with him at his retirement base in Hawaii in August 1997, describes him in a placing on the internet website (<u>owner-biowar-1@afterburner.sonic.net</u>) on 07/02/99 as "author of the pathetically evil paper advocating infectious contraceptives". She goes on to remark:

In the course of the interview, I was struck by the glib manner in which this retired college professor explained his callous plan, and how little he seemed to have considered the possible consequences of what he had advocated. My general impression was of a thoroughly amoral man who was angry at his fellow man and had spitefully advocated the construction of a germ which would sterilize women.

Hall's prescription could be connected with and read as a particularly brilliant contribution to a general policy, set and informed by what has come to be referred to as "The Kissinger Report" alias "NSSM 200" (1974) on the subject of "Implications of Worldwide Population Growth for U.S. Security and Overseas Interests". The said report is in effect a United States presidentially-commissioned in-depth study, analyses and recommendations defining American world population control policy. In the "Executive Summary" of the Report, after reviewing world demographic trends in relation to natural resources, food supplies and economic development since the Second World War, the Report observes:

The political consequences of current population factors in the LDCs (less developed countries) – *rapid growth, internal migration, high percentages of young people, slow improvement in living standards, urban concentrations, and*

*pressures for foreign migration – are damaging to the internal stability and
international relations of countries in whose advancement the U.S. is interested,
thus creating political or even national security problems for the U.S. In a broader
sense, there is a major risk of severe damage to world economic, political, and
ecological systems and, as these systems begin to fail, to our humanitarian values*
(p. 7).

The Report sees this situation as being very critical to U.S.
political and strategic interests and recommends that the U.S., while
taking care not to give the impression to the LDCs of an
industrialized country policy against the LDCs, must assume
leadership in the task of creating conditions conducive to fertility
decline, motivations for smaller family size and developing
alternatives to children as a source of old age security. The Report
goes on to state that a "growing number of experts believe that the
population situation is already more serious and less amenable to
solution through voluntary measures than is generally accepted" (pp.
11-12).

The U.S.A. took effective leadership in the organization of
the first ever World Population Conference (WPC) in Bucharest,
Romania, in August 1974, whose main purpose was to adopt a World
Population Plan of Action (WPPA). At this conference, most of the
so-called LDCs, led by Algeria, backed by several African countries;
Argentina, supported by Uruguay, Brazil and Peru; the Eastern
European countries excepting Romania; but also The People's
Republic of China and the Holy See, expressed their skepticism and
reservations concerning the WPPA. The main thrust of their several
concerns revolved around the view that economic and social
development ought to be considered as a necessary condition for
high fertility reduction, that references to population programmes
involving quantitative and time goals should be de-emphasized and
that national sovereignty should be recognized in setting population
policies and programmes. However, the document finally adopted by
acclamation, with only the Holy See dissenting, took the commitment
to conscious and deliberate control of the world's population to bring
it into harmony with the carrying capacity of the Earth, that is,
making sure that the number of people in the world is such that the
Earth can hold in a sustainable way.

SOUTH AFRICA IN THE APARTHEID ERA

Whether or not there is any credibility to charges of a grand global conspiracy bent on using covert means to decrease the numbers of targeted populations, it is something that has actually been conceived and tried in South Africa during the Apartheid era. After the first ever democratic elections in South Africa in 1994, Nelson Mandela, who walked straight from prison after 27 years to the summit of political power, instituted the Truth and Reconciliation Commission (TRC) as a preliminary condition on which to start building a new South Africa. The commission had the task of listening to testimonies and confessions on gross human rights violations, granting amnesty to perpetrators of politically motivated violence and recommending reparations for victims of these abuses, as a cathartic process that could ease the transition from a radically unjust to a free democratic society. The revelations and evidence that came out of the TRC took many people by complete surprise in their sordidness and callousness. Such, for example, were revelations about the complicity of scientists and health professionals in secret population control projects. In effect, Apartheid South Africa had a secret biological and chemical warfare programme in which the South African Defence Force (SADF) was charged with the development of an anti-fertility vaccine which could clandestinely be used on black people without their knowledge and consent (de Gruchy and Baldwin-Ragaven 1998, p.6).

During the Fourth World Congress of Bioethics (IAB4) in Tokyo, Japan, in November 1998, and the satellite conferences in Tsukuba: TRT4 (Fourth International Tsukuba Bioethics Roundtable) and FAB2 (Second Conference of the International Association on Feminist Approaches to Bioethics) that preceded the congress, many participants listened with total amazement at the eye/ear-witness accounts emanating from the TRC. By far the most remarkable and memorable presentation I listened to on that occasion was that of Wendy Orr, a young, articulate and militant female medical doctor from South Africa who was a member of the TRC under the chairmanship of Archbishop Desmond Tutu. Wendy's vivid account of some of the findings of the TRC was profoundly shocking. She talked about South Africa's Chemical and Biological Warfare (CBW) programme. According to her:

The work of the scientists involved in the CBW programme was covert and secretive in every imaginable way. It was conducted under the guise of private 'front companies' supposedly conducting commercial research and development work; it was funded from undeclared, unaudited secret government coffers; research and experimentation were conducted on a strict 'need to know' basis with little or no communication or interaction between scientists working on different projects or between scientists inside and outside the programme. The aim of the programme was to develop agents harmful to individuals, groups and communities (Orr 1998, pp. 1-2).

Wendy pointed out that, apart from about two or three medical doctors, those involved in the CBW programme were 'chemists, engineers, physicists and veterinarians, who did not see themselves as bound by any particular ethical codes of conduct'. For this reason, she observes that 'present conventional approaches to Bioethics/Medical Ethics may exclude certain groups of scientists or leave gaps which allow them to exclude themselves from ethical obligations.'(p. 2). She further lists a few of the projects that were pursued under the described milieu of secretive and non-interactive scientific research in disregard of all concern for ethics and human rights:

- The production of thousands of kilograms of street drugs (like Ecstasy and Mandrax), supposedly for use as crowd control agents.
- Research into the use of various carrying agents for organophosphate poisoning, for example, beer, whiskey, chocolate, shampoo.
- The development of 'applicators' which were, in effect, murder weapons, for example, screwdrivers which could inject poison into a chosen victim and leave an almost undetectable external puncture wound.
- Research into toxic agents which are easily administered, lethal, tasteless, odourless and undetectable in the body.
- Research into and the stockpiling of millions of drug resistant cholera, anthrax, plague and botulinum organisms.

Most shockingly of all, Wendy revealed that:

- In spite of the fact that South Africa was supposedly subject to international sanctions, the SADF received support for this programme from a number of foreign countries, including the USA, the UK, France, Israel, China and Germany.

Wendy concluded her presentation by drawing attention to the saying that 'all it takes for evil to succeed is for good men (and women) to stand by and do nothing'. She finally wondered what role organizations like the International Association of Bioethics could play in addressing, drawing attention to and halting 'evil' science.

With the foregoing as a background, it goes without saying that any vaccine initiative or other biomedical research in Africa needs to be transparently well-intentioned and unambiguously ethical. Recently, several hundreds of children reportedly died of meningitis in parts of northern Nigeria because of the skepticism of parents against vaccinating their children, on the grounds that the vaccine was contaminated with the HIV virus. The origin of the skepticism is said to be the clinical trial mentioned above in my introduction in which the American company, Pfizer, without bothering about research ethics, allegedly tested an experimental drug in the region some time ago, with disastrous consequences in terms of deaths and irreversible harm to many of the participating subjects.

SOUTH AFRICAN AIDS VACCINE INITIATIVE
Post-Apartheid South Africa has the highest rate of HIV/AIDS infection (currently estimated at 25%) in all of sub-Saharan Africa as well as the most serious current vaccine projects against HIV/AIDS. No other African country possesses better human and material resources to tackle the AIDS epidemic and no other has actually shown more determination, seriousness and self-reliance in doing so than South Africa (Galloway 1999). Other African countries such as Uganda, Kenya, Ghana, Ivory Coast, Tanzania, have mainly been looking outward towards the Western world and are involved in collaborative efforts whereby they provide sites to Western researchers and sponsors for the testing of candidate vaccines, in the hope of somehow receiving preferential treatment, in case a successful vaccine is found.

The South African AIDS Vaccine Initiative (SAAVI) was established in 1999 and charged with, among other things, producing a safe, effective, affordable and accessible local vaccine for the South African population by the year 2005. SAAVI collaborates with the International Aids Vaccine Initiative (IAVI) that was founded in 1996 with the objective of facilitating the development of a suitable HIV vaccine that could be used throughout the world. SAAVI is also a partner with some Western projects based in the USA that are attempting to develop a candidate vaccine using gene technology. SAAVI is a South African national initiative, sponsored by the South African government, and coordinated by the Medical Research Council (MRC) of South Africa. It is a semi-public, semi-private outfit, funded from both government and non-governmental sources at home and abroad. The efforts of SAAVI are very much on course but it cannot be said at this point whether it will succeed in its aim of producing a safe, cheap, effective and affordable vaccine for South Africa by the targeted year of 2005. What can, perhaps, already be said at this point is that such a vaccine, resulting from a self-reliant local initiative, holds the best hope of defeating the HIV/AIDS monster in South Africa and, indeed, in the rest of sub-Saharan Africa.

Recent and ongoing experiences related to the patenting, licensing and marketing dispute, pitting the Pharmaceutical Manufacturing Association (a consortium of about 40 Western pharmaceutical companies) against some African governments, philanthropic international NGOs and AIDS activists show that poor developing countries, particularly in sub-Saharan Africa, would be very naïve indeed to hang their entire hopes of ever defeating HIV/AIDS on Western researchers and their sponsors (including those who from time to time announce drastic discounts in the cost of their drugs – a standard marketing strategy). The cost of niverapine, for example, has been reduced by an alleged 75%, but it is still unaffordable for the vast majority of African nursing mothers. The South African government and the international NGO, Médicines Sans Frontirères (MSF), are learning hard lessons in their efforts to profit from an offer from places like India and Thailand of cheaper essential HIV palliative medicines (antiretroviral cocktail drugs) to make such medicines more accessible to poor South Africans. A similar scenario will be enacted in any African country that contemplates the same or similar way out of its predicament *vis-à-vis* the HIV/AIDS epidemic. And while the patent disputes will be ragging and dragging on, in or out of the courts, patients will continue dying in their numbers.

TRADITIONAL AFRICAN ATTITUDES TOWARD DISEASE AND COMMERCE

The central importance of commerce in Western thought and practice and the linkage of medicine and healthcare to the market in the Western system (Stoline & Weiner 1988), though it is not without its positive side and advantages, makes it inevitable that Western industries and business companies have an understandably incorrigible outlook and attitude, motivated and driven by pure economic logic and the lure of profit. This attitude and outlook is always on the lookout for and sees economic opportunities and advantages in even the most distressing of human calamities. It would be quite vain to hope that such a situation could be changed by mere wishful thinking or, under any circumstances, in the short run. Many Westerners who are themselves unhappy with this state of affairs are quite impotent to do anything about it. In Africa, where traditional attitudes to both commerce and to disease and treatment are quite different, self-reliant initiatives, supported by genuine humanitarian international help, are the surest way forward against the AIDS and similar epidemics.

The traditional African outlook and attitudes to both commerce and disease are worth briefly remarking here. They are succinctly captured in my mother's remark of a few years ago, to the effect that medicine and commerce are bad companions (Tangwa 2001, p. 37). In traditional Africa, curative medicine particularly but also diagnostic and prophylactic medicine to some extent, were completely divorced from commerce. The resources of the immediate community were always mobilized on behalf of a seriously sick person and no traditional healer worth his name could directly charge any fees for his/her services, under pain of being considered a quack or even a con person or of losing his/her special skills and endowments (Tangwa 1999, p. 277). Any typical African market and the system of haggling that goes on there is very indicative of an attitude based on a deeply ingrained value system and worldview. In an African market the prices of goods are never fixed. The price quoted to a prospective buyer by any seller usually depends on who the former happens to be and the quoted price, in any case, is always an invitation for an animated and lively dialogue in which both the buyer and seller reveal and learn information about each other and his/her particular situation and circumstances. I have once witnessed a scene in an African market (the weekly Mbvë Market of Kumbo Town in Cameroon) where, on being told the price of an article, the buyer, without any further ado, immediately removed the amount of money quoted, threw it in front of the seller, took the article and started walking hurriedly away. The seller got worried enough at such comportment to the extent of running after him, giving him back his money and retrieving the article, under the pretext that it was, after all, not for sale. The real reason for such reaction was that he was worried that the buyer may not be a genuine human being and that his money and manner were liable to 'bring one misfortune (*lon*)'.

In the course of haggling over an article, the seller and buyer sometimes discover that they are blood relations, in which case, the latter could end up getting the said article for free or with a generous addition or other gift. And whenever the product needed is for a patient or is an ingredient for preparing the medicine of a patient, then, invariably, it is given either freely or very cheaply. Always, the poverty or otherwise of both the buyer seller and his/her situation and need are a factor to be considered in reaching the final price of any product or article, in any typical African market. The same attitude underlay the traditional practice whereby it was permissible for anybody to pick fallen fruits from any orchard, no matter to whom the trees belonged, or to harvest enough of a directly edible crop to stem great hunger for the moment. As long as you did not climb up the fruit tree or harvest the crop to store or resale (both of which were strictly taboo), you were considered to be well within the narrow bounds of morally permissible action.

The general underlying idea was that no one should die from sheer need or want in a situation where the remedy was readily available. Note that this is the complete antithesis of the modern Western economic idea and practice, whereby the more desperately you need a product or a service, the more you are required to pay for it under the so-called law of demand and supply. For sure, the African traditional practices and the world-view and attitudes that sustained them had their negative aspects and disadvantages and, moreover, they now seem irretrievably to be receding out of history, but some of the spirit and moral imperatives that imbued them could still be salvaged, modernized and even globalized. It would be a good idea to attempt confronting in a more sustained and systematic way the antithetical moral intuitions and attitudes of traditional Africa with those of modern Western society which have been exported and imported all over the globe, but this is not the time or place to do it.

VANHIVAX IN CAMEROON

A local initiative in Cameroon (Tangwa 2002, pp. 222-225) worth remarking is that of Professor Victor Anomah Ngu, an oncologist and former laureate of the Lasker award for cancer research (1974). Professor Ngu, a former minister of health and former Vice-Chancellor of the University of Yaounde, started his research for a vaccine against the HIV/AIDS virus more than a decade ago, with inspiration from his life-long work against cancer. His seminal hypothesis, published in *Medical Hypotheses* (Ngu 1994 and Ngu 1997) is that, being an 'enveloped' virus, the HIV is 'perceived' by the host immune system as 'partly-self' because of the presence of host cell wall membrane on the viral envelope and that this situation leads to an ineffective response by the body's immune system to the virus. He reasoned that viral core antigens *without the envelope* would be 'perceived' as 'non-self' by the host immune system, thereby eliciting an appropriate and effective immune response. In a normal uninfected person, therefore, such 'unenveloped' core antigens, he claims, would serve as an effective vaccine.

As a non-medical person with more than a passing interest in science and medicine, I found Professor Ngu's hypothesis reasonable, coherent, interesting and well worth a serious investigation in the face of the grave threat that is HIV/AIDS. But when he presented his hypothesis and preliminary investigative findings to his medical colleagues (many of them involved by government accreditation in 'the fight against HIV/AIDS') and to appropriate international agencies, with a view to obtaining assistance in systematically testing his hypothesis, he at first met with only skepticism and even derision. In that situation, he then proceeded, with meager personal resources, to demonstrate his claims, in sero-positve patients, by preparing an *autologuous* vaccine for each of the patients. The results he obtained, even with his rudimentary facilities, clearly seemed to confirm his hypothesis.

In September 2000, Professor Ngu and I presented a joint paper at the Fifth World Congress of Bioethics, Imperial College, London, UK, under the title: "Effective Vaccine against and Immunotherapy of the HIV: Scientific Report and Ethical Considerations from Cameroon" (Ngu and Tangwa 2000). The gist of our joint paper can be summarized in eight points, as follows:

- The HIV/AIDS is a terrifying and deadly disease, particularly for sub-Saharan Africa, where it has already claimed more than 17 million victims in less than two decades and is poised to wipe out the entire population.
- The central problem of the HIV infection is that immune responses by the body fail to kill the virus.
- The reason for this failure is that it is an enveloped virus. The envelope of the HIV is derived from host CD4 cell wall and killing the virus with its envelope will also kill host CD4 cells – leading to an auto immune disease. The virus has used this threat to blackmail and to block, so to speak, effective immune responses that alone can kill the virus.
- The ideal solution for the body is to provoke immune responses to the HIV that is without an envelope. Such new immune responses will effectively kill only the virus, sparing the viral envelope and CD4.
- To obtain such new effective immune responses in an uninfected person, one starts with HIV antigens from which the envelope has been destroyed before hand. Such antigens, which cannot infect, constitute an effective vaccine for the HIV. All HIV antigens, with only the envelope destroyed, constitute an effective HIV vaccine.
- Being unable to have such a vaccine tested on normal healthy persons, we tested it in patients, using their own auto-vaccine, under special conditions and have so far obtained results that confirm our hypothesis.

- This test showed that the auto-vaccine provoked immune responses that kill the virus, confirming its effectiveness as a vaccine. This procedure of using vaccines to induce the killing of the virus in patients constitutes a form of immuno-therapy for the HIV. A vaccine produced on the same basis as the auto-vaccine will be acceptable for trial on the public, because it has proved its worth in patients.

- In our humble opinion, such a vaccine, in a world in which medicine and commerce seem to have become inseparable companions, holds the best hope of stopping the AIDS epidemic, especially in sub-Saharan Africa, because, *inter alia,* it is completely effective, simple and cheap to produce, using traditional as against modern high technology methods, and therefore affordable for the poor masses, who are highly vulnerable and mostly affected by the virus.

Those who listened to our London presentation of what is in effect a candidate vaccine to which the name VANHIVAX has been given, seemed completely taken by surprise. First of all, no one seemed to have entertained the possibility of a 'homegrown' candidate vaccine proposal coming from Africa. What seemed to have been expected were proposals as to how Western researchers could carry out ethical and acceptable HIV/AIDS vaccine research in Africa, with or without the collaboration of local scientists. Secondly, the surprise was expressed as to why, with the interesting clinical results presented, no Western researchers or funding agencies had shown interest in VANHIVAX.

Ever since, Professor Ngu has obtained even more convincing results but VANHIVAX has not yet aroused the sort of interest one would expect in the prevailing situation and there are those both at home and abroad who have continued to dismiss his work as being 'unscientific' without bothering to try to demonstrate such dismissive opinion in a scientific manner. Others, while admitting that he may have discovered 'some sort of cure for AIDS, perhaps', seriously contest his claim to having a protocol for an HIV/AIDS candidate vaccine, on the grounds that a vaccine, by definition, is not something you give to a patient (which fulfills the definition of a drug) but rather something you give to a healthy person to prevent infection. This, of course, is trying to argue against a substantive claim by means of a stipulative definition. If the candidate vaccine happens to have immuno-therapeutic action/effect, why should it be excluded from the category of vaccines? A boxer who happens to be also a sprinter can not be excluded from the category of boxers on the grounds that boxing does not involve sprinting.

A possible explanation for the lack of interest and dismissive attitude is that Western researchers and sponsors of HIV/AIDS vaccine research are too fascinated with 'state of the art' high technology candidate vaccines in which, moreover, they have invested enormous amounts of intellectual and physical energy, money and other material resources, whereas VANHIVAX will require a simple traditional technology which would not arouse much economic interest, if any at all. As for other African biomedical researchers and health policy authorities, many of them seem to be already involved, in one way or another, in collaborative projects with Western agencies or counterparts, which they find intellectually, politically, professionally or financially fascinating. It seems very difficult to come out from this frame of mind and to pay adequate attention to a simple proposal based on basic fundamental science and requiring only outmoded technology and modest financial stakes.

CONCLUSION

The Western approach to disease and illness and to the AIDS pandemic in particular, like many other things Western, is overly analytic, statistical and business-like. This business-like statistical analyticity may, from some perspective, appear like the epitome of rationality, but it ignores other perspectives and other aspects of living and being human that ought to be taken into account. The analytic paradigm of knowledge is, of course, very important and is perhaps the most important kind of knowledge, present to varying degrees in all human cultures, but it is not the only one. There are other types of human knowledge such as the intuitive, the metaphoric, the parabolic, the mythic, etc., which are, more or less, much better developed and more prominent in non-Western cultures.

Globalization should not be allowed to install analytic knowledge as the sole paradigm of knowledge because there are aspects of reality and human life and existence with which that paradigm, important though it certainly is, cannot adequately deal. Commerce and the profit motive, though important and indispensable in a world of limited resources, should not be allowed to invade all aspects of life and all vital resources for living. The HIV/AIDS epidemic and other life-threatening or endemic diseases in Africa are not likely to be easily controlled or eradicated by the present over-emphasis on study, statistical collection and analysis of data. It is not simply knowledge about how many people are being infected with HIV every second or how many children are dying from AIDS or other diseases every minute that will turn the situation around. In African attitudes and expectation, when people are ill they want to be helped, not studied and analyzed, and when people know that there is no cure for their illness, they quietly accept their fate with courage and hope. Some cultural customs, practices and attitudes may be a stumbling block to the control or eradication of certain diseases, and they need to change. But cultural customs, practices and attitudes change only slowly over long periods of time, and this must be taken into consideration, along with the thought that the primary vocation and duty of the healer and of medicine in general is to offer relief from pain and suffering, not to change people's way of life.

The dominant Western world-view mainly sees human beings as autonomous individuals or groups of such individuals, each competing with the others in a struggle to maximize own advantages. Such ideology is correctly described as 'possessive individualism'. This view, together with the principle that 'might is right' are what have ruled the world since David Livingstone first set foot on the African continent. The general attitude engendered by these two principles is what led to colonialism, the slave trade, and caused the first and second world wars, from which the principles came out even stronger and better reinforced in human consciousness. But the attitude is today clearly on the decline, especially in those parts of the world where it has held sway and been most influential. Taken together with the positive possibilities of the ongoing process of globalization, this fact offers the best chance that human beings have ever had for creating a better world for all inhabitants of the earth. A global communitarian outlook, borrowed from African cultures and appropriately enlarged and modernized, could set the prescriptive standards and goals of our impending global village.

The dilemmas confronting African countries and the international community in the face of the HIV/AIDS and other life-threatening pandemics might be put in the following parable.

Suppose that you are very poor and almost starving. You are eating the last few rotting yams left in your barn. In this situation, it would be highly morally commendable if you were to offer some of your rotting yams to a neighbor who is equally starving but whose barn is already completely empty. But, suppose that I, your third neighbor, have a barn full of yams from which I regularly select rotting yams to throw away. I now select my rotting yams and, instead of throwing them away, I offer them to you, cheaply or even freely. You will surely be very grateful and there is no doubt that my action is beneficial to you, in some sense. But is my action morally worthy or commendable? Is my admittedly beneficial action the type which could possibly set the standard and limits of moral action and behavior in an interdependent global community? What if I further suggest that your ultimate problem is lack of agricultural know-how and that the ultimate solution to your problem is to hand over all your ancestral lands to me for proper exploitation, abandon your native ways, and listen and look up to me henceforth?

We should be very careful not to confuse prudential calculations, commerce or economic considerations in general, important as they may be in their appropriate context, with the ethics of healthcare or HIV/AIDS vaccine research in particular. Ethics is not about self-interest, not about bargaining, not about realism of the present moment, not about learning new useful habits, but about what ought/ought not to be done in a given situation, anywhere at any time. It is also about sympathizing and empathizing with fellow humans in need, doing unto them as we ourselves would be done by, treating them as ends in themselves and never as mere means to any other end and, above all, doing them no harm (*primum non nocere! Above all do no harm!*).

REFERENCES

Benatar, S. R. (2000): Avoiding Exploitation in Clinical Research. *Cambridge Quarterly of Healthcare Ethics.* 9: 562-565.

Benatar, S. R. and Singer, P. A. (2000): A New Look at International Research Ethics. *BMJ.* 321: 824-565.

Cantwell, A. (1988): *AIDS & The Doctors of Death: An Inquiry Into the Origins of the AIDS Epidemic.* Los Angeles: Aries rising Press.

------------ (1993). *Queer Blood: The Secret AIDS Genocide Plot.* Los Angeles: Aries Rising Press.

De Gruchy, J. and Baldwin-Ragaven, L. (1998): Population Control: Health Professional Accountability in Apartheid South Africa...and Beyond. Unpublished Manuscript, 1-15.

Galloway, M. (1999): Vaccine Initiative on Track as First Funding Allocations are Made: Four Projects to Receive Funding. *AIDS Bulletin*, 8(4): 10-11.

Global Forum for Health Research (Helping correct the 10/90 gap): The 10/90 Report on Health Research 2001/2002.

Hall, John B. (1996): Negative Population Growth: Why We Must, and How We Could Achieve It. *Population and Environment*. 18 (1). Diffused on Internet List serve: (afri-phil list), 31/01/1999.

Horowitz, L. G. (1996): *Emerging Viruses: AIDS & Ebola.* Rockport MA: Tetrahedron Publishing Group.

Human, D. (2000): Editorial' *WORLD MEDICAL ASSOCIATION DECLARATION OF HELSINKI*, Unpublished Text.

Kissinger, H. (1974): *National Security Memorandum: NSSM 2000 (THE KISSINGER REPORT).* April 24.

Luinggvist, K. I.(1992): *AIDS Tabu.* Stockholm: Carlssons Bokforlag.

Macklin, R. (2003): Bioethics, Vulnerability and Ethics. *Bioethics* (forthcoming).

Ngu, V. A. (1994): Chronic Infections from the Perspective of Evolution: a Hypothesis. *Medical Hypotheses.* 42: 81-88.

------------ (1997). The Viral Envelope in the Evolution of HIV: a Hypothetical Approach to Inducing an Effective Immune Response to the Virus. *Medical Hypotheses,* 48: 517-521.

Ngu, V. A. and Tangwa, G. B. (2000): Effective Vaccine Against and Immunotherapy of the HIV: Scientific Report and Ethical Considerations from Cameroon. Unpublished Manuscript, 1-8.

Orr, W. (1998): Can "Good" Ethics Be Applied to "Evil" Science?' (Unpublished conference presentation).

Sabatier, R. (1988): *BLAMING OTHERS: Prejudice, Race and Worldwide AID.* Tinker, J. (ed.). London, Paris, Washington: The Panos Institute.

Stoline, A. M. and Weiner, J. P. (1988): *The New Medical Marketplace: A Physician's Guide to the Health Care System in the 1990s.* Baltimore and London: The Johns Hopkins University Press.

Tangwa, G. B. (1996): Bioethics: An African Perspective. *Bioethics.* 10(3): 183-200.

------------- (1999). GENETIC INFORMATION: Questions and Worries from an African Background. In Thompson, K. A. and Chadwick, R. (eds), *Genetic Information: Acquisition, Access and Control.* New York, Boston, Dordrecht, London, Moscow: Kluwer Academic / Plenum Publishers.

------------- (2000): Traditional African Perception of a Person: Some Implications for Bioethics. *Hastings Center Report.* 30(5): 39-43.

------------- (2001): Traditional African Perception of a Person: Some Implications for Health Research and Health Care Ethics. In Rugemalila, J. B. and Kilama, W. L. (Guest Eds), *ACTA TROPICA, Supplement Proceedings Seminar on Health Research Ethics in Africa.* Amsterdam - London - New York – Oxford – Paris – Shannon – Tokyo: ELSEVIER, pp. S32-S39.

------------- (2002): The HIV/AIDS Pandemic, African Traditional Values and the Search for a Vaccine in Africa. *Journal of Medicine and Philosophy* 27(2): 217-230.

Wiredu, K. (1994): Philosophy, Humankind and the Environment. In Oruka, O. (ed.), *Philosophy, Humanity and Ecology: Philosophy of Nature and Environmental Ethics.* Nairobi, Kenya: ACTS Press.

INTERNATIONAL GUIDELINES ON MEDICAL RESEARCH/PRACTICE

Nuremberg Code (1947)

Declaration of Geneva (Physician's Oath) (WMA 1948, 1994)

Declaration of Helsinki (WMA 1964, 1975, 1983, 1989, 2000)

International Guidelines for Ethical Review of Epidemiological Studies (CIOMS 1991)

International Ethical Guidelines for Biomedical Research Involving Human Subjects (CIOMS 1993, 2002)

Note for Guidance on Good Clinical Practice (ICH 1996)

Universal Declaration on the Human Genome and Human Rights (UNESCO 1997)

Convention for the Protection of Human Rights and Dignity of the Human Being with Regard to the Application of Biology and Medicine: Convention on Human Rights and Biomedicine (Council of Europe 1997)

The Ethics of Clinical Research in Developing Countries (Nuffield Council on Bioethics, Discussion Paper 1999)

Operational Guidelines for Ethics Committees that Review Biomedical Research (WHO 2000)

Ethical Considerations in HIV Preventive Vaccine Research (UNAIDS 2000)

Ethical and Policy Issues in International Research: Clinical Trials in Developing Countries (National Bioethics Advisory Commission – NBAC – 2001)

The Ethics of Research Related to Healthcare in Developing Countries (Nufflied Council on Bioethics, Launch of New Report 2002)

CHAPTER THIRTEEN

IS BIOETHICS LOVE OF LIFE?
AN AFRICAN VIEW- POINT.

[This chapter is published in one of the Newsletters of the International Association of Bioethics (IAB)]

In his recent book, *BIOETHICS IS LOVE OF LIFE: An Alternative Textbook*, (Eubios Ethics Institute, 1998), Darryl Macer suggests that 'love of life' is 'the simplest and most all encompassing definition of Bioethics'. Macer does make a very good case for his way of looking at things. But an African who thoroughly agrees with him is more likely to say that Bioethics is like love of life rather than that it is love of life. The resort to similes, metaphors, proverbs, parables, etc. by Africans in expressing views, beliefs, convictions, etc., has deep epistemological and didactic foundations.

I would prefer to say that Bioethics is RESPECT or REVERENCE FOR life rather than LOVE OF life. Love seems to me to be too complex, generic and diffuse a concept, and one with too many problematic associations and connotations, to conveniently and economically carry our characterisation of Bioethics. What I am looking for in my attempt to describe Bioethics is a concept that would accommodate at one and the same time my love of crickets (not the English game!) and termits, my indifference to glow-worms and butterflies, my fear of snakes and scorpions and my aversion towards vultures and cameleons. It seems to me that RESPECT and REVERENCE would do this better than LOVE even though it could be argued that the former are included in the latter.

In global Bioethics, however, it is less important to agree on a description of Bioethics than, what is much more important, on the ethical concerns and imperatives that underlie it.

I just made allusion to my love, aversion and indifference, respectively, to certain biological entities. This reminds me of an experience I had with snails in Nigeria. Snails are strictly taboo among my people, the Nso' of the grassy highlands of Bamenda in North-western Cameroon. However, once, while travelling by bus from Ibadan to Benin city in Nigeria, I found myself sitting directly opposite a young lady who was so beautiful and seemed so unconscious of this fact that I had a strong impulse to kneel down and adore. I could hardly take my eyes off her and, when, at some point along our journey, she brought out from her hand bag some fried snails and plantains and started eating, she must have mistaken my enchanted and fascinated gaze as directed to her snack, for she offered me a good portion of it and in such a charming manner that refusal was out of the question. Now when I tasted the snack, I was pleasantly surprised that snails, towards which I felt a certain cultural horror, in fact, tasted so good. While back home on holidays, I related my experience with snails in Nigeria to my family members, and I could see everyone instinctively recoiling away from me as I told the story, until I added that I had met people who were shocked and horrified when I told them that in Nso' we greatly cherish termites, crickets and certain grasshoppers as delicacies. All this has lessons for the attitude and disposition we require in seriously discussing any issue cross-culturally.

There is a further reason why I would not want to characterise Bioethics as 'love of life'. In Lamnso' there is a distinction between love in general (kong) and reciprocal love (kikongnin). The former, which is both noun and verb, and which also means 'will', 'like' etc. is rather passive while the latter is active and can occur only between equals. When we say that God loves us and that we love God, we mean love in the sense of kong. No human being can have a relationship of kikongnin with God. Human beings can love (kong) animals, plants and inanimate objects, but none of these can enter into a relationship of kikongnin with any human being. Now while kong can be unconditional, kikongnin is not possible without reciprocity.

Nso' morality is human centred in the sense that only human beings are deemed to be subject to it but not in the sense that moral concerns are limited to human beings. Human beings have moral obligations, responsibilities, duties, etc. towards God, non-human animals, plants and inanimate nature and things, but these are not considered as being in any sense reciprocal.

The idea that seems viable and operative in covering both cases of spontaneous love and love as a moral duty or obligation, in covering both kong and kikongnin, is that of reverence or respect. Reverence has the further advantage of not being too all encompassing as a concept and of not having too many misleading associations or connotations.

If the first images that come to someone's mind when s/he hears the word 'love' are those of Monica and Bill or of a paedophile, unable or unwilling to 'make love' to a fully grown mature adult, but quite able and willing not only to do so to a week old baby but even to video-record this 'love' for the benefit of other 'child-lovers' around the world, such a person would surely need a rather complex spiral flight from there to the idea of Bioethics.

A friend of mine with whom I share the same cultural background, Bongasu Tanla Kishani, has a collection of poems to which he has given the title *Kong-la-njoh*, meaning **love without ill-fortune**. Another Cameroonian, Ngongwiko, has a novel entitled *Taboo Love*. Some types of love are liable to bring misfortune or disaster. That is why in Nso' we have taboos against certain things including certain types of love such as that of tampering with the immature (plants, animals and humans). The violation of such taboos brings misfortune and disaster, not only on the violator but on the earth, plants, animals and humans indiscriminately; it endangers the collective well-being or existence of all biological beings and requires ritual cleansing and purification as an antidote.

Peaceful coexistence, promoted by the idea of **live and let live** seems to me to be central to Bioethics. Such peaceful coexistence is possible even where love in the active sense is absent, but it would not seem possible where reverence and respect are absent; hence, my preference for defining Bioethics as reverence and respect for life.

CHAPTER FOURTEEN

GLOBALISATION OR WESTERNISATION? ETHICAL CONCERNS IN THE WHOLE BIO-BUSINESS

[This chapter was first delivered as a conference presentation at the Fourth World Congress of Bioethics, Tokyo, 4-7 November, 1998, subsequently published in **Bioethics,** *Vol. 13, Nos. 3 and 4, pp. 218-226, July 1999, and reprinted in* **Ethical Health Care,** *edited by Patricia Illingworth & Wendy E. Parmet, New Jersey: Upper Saddle River, 2006, pp. 471-476]*

ABSTRACT

Increasing awareness of the importance of the biodiversity of the whole global biosphere has led to further awareness that the problems which arise in connection with preservation and exploitation of our planet's biodiversity are best tackled from a global perspective. The 'Biodiversity Convention' and the 'Human Genome Project' are some of the concrete attempts at such globalization. But, while these efforts are certainly very good at the intentional level and on paper, there is, at the practical level of implementation, the danger that globalization may simply translate into westernization, given the Western world's dominance and will to dominate the rest of the globe. How is 'global bioethics' to be possible in a world inhabited by different cultural groups whose material situation, powers, ideas, experiences and attitudes differ rather markedly and who are not, in any case, equally represented in globalization efforts and fora? One index of the pertinence of this question is that talk about biodiversity, biotechnology, biotrade etc. is being increasingly matched by talk about biopiracy, biorade, biocolonialism, etc. In this paper, I attempt to explore and develop these very general concerns.

GLOBALISATION OR WESTERNISATION? ETHICAL CONCERNS IN THE WHOLE BIO-BUSINESS

A handshake should not go beyond the elbow (Chinua Achebe)

INTRODUCTION

The above African proverb is the convenient nail in conceptual space, as it were, on which I would like to hang the spirit, as distinguished from the mere letter, of what I would like to say here. I will not, of course, attempt to interpret the proverb because another proverb says that 'it is the fool who says a proverb and then proceeds to interpret it himself/herself' or, alternatively, that 'it is only to the fool that when you say a proverb you also need to interpret it' This African proverbial approach will help me in avoiding too much bluntness which often tends to turn dialogue into confrontation. Furthermore, since Western space and time worship will surely not permit me to say all I could have said here, I would like to commend rather my figures of speech and not the speech itself to your further reflection.

GLOBALISATION

Globalization, as a descriptive process, has been made possible and inevitable by advances in science and technology, especially in loco-motion and communication technologies. The net result of these advances has been increased contact between the various peoples and cultures that populate the world. Thanks to this state of affairs, the world is today, unlike yesterday, aptly described as a 'global village'. This villagisation of the world should have as one of its logical consequences the slow but sure transformation of the world into a 'rainbow village', by analogy with our appellation of South Africa, in our optimistic moments, as the 'Rainbow Nation'. Resistance to this aspect of the process of globalization, exemplified in the savagery with which persons from some parts of the globe are sometimes forcibly excluded from some other parts, cannot but create a lot of tension within the process. Modern technology, in general, and locomotion and communication technologies, in particular, are, of course, inventions of the Western world which have been very effectively used, *inter alia*, in colonizing and dominating peoples in other parts of the world.

Globalization, as a prescriptive process, arises from increasing awareness of both the diversity as well as interdependence of the various parts, peoples and cultures of the world. Globalization in this sense, is essentially a moral concept. Underlying such blueprints of globalization as the *Biodiversity Convention* and the *Human Genome Project*, are clear ethical impulses, concerns and imperatives. But between globalization as a descriptive process and globalization as a prescriptive ideal, there is a difference which involves the danger that globalization might end up as or, in fact, might not and never has been more than, mere Westernization, given the history and reality of Western industrial-technological power, colonization of non-Westerners, domination and insensitivity to all things non-Western.

WESTERNISATION

The European **Industrial Revolution** of the 18th and 19th centuries drew its impetus from the slogan that 'knowledge is power' convertible into commercial value, from the idea that all knowledge is unqualifiedly good, from the belief that nature is, in principle at least, completely knowable and controllable, and from perception of the universe as something which ought to be explored, subdued, dominated and exploited. These ideas and the attitudes they created had their origins and foundation in Judaeo-Christianity and its remarkable creation myths, but they led, paradoxically, to secularization, desacralization and profanation of everything in the universe - the very antithesis of the Judaeo-Christian ethic. The spirit of omnivorous discovery which the **Industrial Revolution** engendered and made possible in Europeans guided them to all parts of the globe where they discovered peoples and cultures so different from theirs that they felt reluctant to qualify them as 'human'. From then on, Europeanization (Westernization) of other peoples and cultures appeared naturally in their eyes as humanization and civilization. It is in this way that both altruistic and egoistic motives became mixed and confounded in the relationship between the technologically very advanced Western world, peoples and culture and other (technologically less advanced) worlds, peoples and cultures. Since the Industrial Revolution, technology has been propelled to great heights by Western commerce and the profit motive, by war and the will to dominate, by pure epistemological and scientific curiosity, as well as (occasionally) by the altruistic urge to improve human well-being. In this process, Western culture has developed the penchant for patenting, monopolizing and commercializing any of its so-called discoveries and a nach for spreading and promoting its ideas, vision, convictions and practices under the guise of universal imperatives of either rationality or morality which ought to be binding on all human beings who are sufficiently rational and moral. I have metaphorically described Western culture as having a big mouth but small ears (Tangwa, 1996, p. 185).

MORALITY

In spite of the remarkable pluralism and cultural diversity of the world, core ethical values are evidently the same for all human beings, if these are carefully separated from and not confused with mere customs, cultural practices, preferences, idiosyncrasies and positive laws. Law and custom are evidently subordinate to morality in the sense that they are ultimately to be justified by reference to morality but not vice versa. A morally unjustifiable law or custom ought not to exist whereas a moral or ethical principle cannot be justified or unjustified by reference to either law or custom. Moreover, it is in the nature of laws and customs to differ from place to place, from culture to culture and even from time to time within the same place and culture, whereas such a situation would be intolerable in morality, properly considered. Universalisability, as Kant and other Western moralists have so rightly argued, is a necessary and indispensable condition of morality. We recognize moral rules from their unconditional imperativeness (or imperative unconditionality). This does not, of course, mean that there cannot be justifiable exceptions to moral rules but only that such putative exceptions do not reverse the rule or, in any way, affect the validity of its imperativeness and unconditionality. (Wiredu, 1995, p.39). 'Moral relativism', in my view, can, therefore, only arise from somehow conflating or confusing customs, laws and practices with morality, strictly considered. Morality can and should overrule both law, custom and practice; in fact, morality can and should overrule every other thing except its own very rationale.

A GOOD WILL

From the subjective and practical point of view, the first and most important principle of morality is what in Lamnso' is called *shiliv she jung shi* (a **good will**, literally, a good heart). If we always act out of good will, we can never be morally blame-worthy. However, a good will is only a necessary but not a sufficient condition for moral rightness. A subjectively good will is putatively quite compatible with objective moral error. This has the consequence that one may act from a morally worthy motive and still be morally wrong - subjectively right but objectively wrong.

This apparently contradictory situation is due to human epistemological limitations. A good will excludes the possibility of knowingly and willingly doing wrong. And, if all moral evil knowingly and willingly done were to be excluded from the sum-total of evil in the world, there would surely be very little left. Kant was also quite right, at least, in considering a good will to be central to all morality. Kant derived his famous doctrine of the **categorical imperative** from the notion of a good will. In the opening sentence of his *Groundwork of the Metaphysic of Morals*, Kant declares:

It is impossible to conceive of anything in the world, or even out of it, which can be taken as good without qualification, save only a good will.

Willing is, of course, something subjective and there is, therefore, an important subjective element or component to morality, captured in the idea and importance of **conscience**, which does not, however, exclude inter-subjectivity or the possibility of objective appraisal of what is or is not moral.

All this has the consequence that there is nothing in the world, or even out of it, which is superior to morality. And it is in this light that we can consider positive laws, customs and taboos as derivatives of and aids to morality, even though they are liable to be conflated and confused with it.

BIOETHICS

Bioethics, as an area of particular concern, is a recent specialized branch of morality, whose focus is the dilemmas, questions, worries and controversies arising from modern Western medicine, biomedical research, bioengineering, biotechnology and attendant/allied processes. Of course, science and technology, in general, have always raised ethical problems and there is a sense in which Bioethics has always been present in all human societies at all times. Professor Van Rensselaer Potter of the University of Wisconsin-Madison, USA, claims to have coined the word 'bioethics' in 1970. (Potter, 1996, p. 2). By it he intended an extension of ethics to cover not only medical ethics but environmental and agricultural ethics, in short, 'the application of ethics to all of life' (ibid). He complains that the medical profession has seized upon the term 'bioethics' and tried to restrict it to medical ethics. Thus, in a broader sense, bioethics can be considered as covering all possible ethical problems that arise or may arise, not only from or within the biological sciences, but in relation to life and living things generally. However, bioethics, in the more restricted sense, is of special concern, because of the potential of biotechnology to transform our lives and our world and because of the rapidity, novelty and magnitude of the problems that it raises.

BIODIVERSITY

The biodiversity of the world, which is similar to its cultural and linguistic diversity, is a gift, so to speak, of evolution and certainly a value which we need only to recognize, accept, use and protect. The Darwinian theory of evolution, with its key concepts of *natural selection* through *survival of the fittest*, may go a long way in explaining biodiversity as a scientific fact, but biodiversity cannot be completely explained and remains, partly, one of the mysteries of nature. Natural gene mutation and natural selection through survival of the fittest, can be considered as nature's own version of biotechnology, with an in-built safety mechanism for nature as a whole. This 'natural biotechnology', as it were, is, by its very nature, gradual and piecemeal, always allowing any ecosystem, including soil, plants, animals and humans, the chance to adjust as painlessly and imperceptibly as possible to changes. Human biotechnology, on the other hand, lacks this advantage and, therefore, ought to proceed more cautiously, always attempting in advance, as far as possible, to assess carefully its short, medium and long-term impact and consequences. Such assessment ought to include the possible cumulative effects on the socio-economic and cultural practices and way of life of all communities likely to be affected by any putative technological intervention and should, in any case, never be limited merely to considerations of marketability and profitability. Failure to do this would be tantamount to using biotechnology in a reckless and ethically unjustifiable way.

GLOBALISATION AND WESTERNISATION

Some of the earth's most valuable wealth, in the form of remarkable variety of biological entities and forms, is to be found in the technologically relatively innocent parts of the earth. Africa, for instance, presents remarkable diversity and variety, not only ecologically and culturally, but biologically. This diversity and variety is what has shaped and coloured African social systems, ethical, religious and metaphysical ideas. The African metaphysical world-view can be aptly described as eco-bio-communitarian (Tangwa, 196, p.192), implying the recognition and acceptance of interdependence and peaceful coexistence between the earth, plants, animals and human beings. This world-view can be contrasted with its Western counterpart which can be described as individualistic and anthropocentric. Within the African outlook, humans are more humble and more cautious, epistemologically more sceptical of their own capabilities and, therefore, practically more conciliatory and respectful of other people, animals, plants, inanimate things and the invisible/intangible forces of nature. They are, thus, also more timorous of tampering with nature and more disposed towards an attitude of *live and let live*.

If this background is not taken into consideration in 'globalizing' Western technologies, then globalization may be simply co-terminus with Westernization. It would be like a handshake that has gone beyond the elbow. Africans are no strangers to the experience of losing a whole arm in what they went into believing to be a friendly handshake. Technology, in itself, has nothing to do with what people think or believe how they conceive of or organize their lives. The globalization of Western technology should not be accompanied by the globalization of Western ways of thinking and acting, Western ways, manners and style of doing things, Western idiosyncrasies and eccentricities. Other cultures should be able to beg, borrow or buy Western technology without having to take it along with all its Western packaging, its entire surrounding value system.

If globalization is to be meaningful and justifiable, in a world populated by different cultural groups, whose respective material conditions, level of technological development, powers, ideas, experiences and attitudes differ rather widely, then it must be premised on certain pertinent presuppositions. Some of such pertinent presuppositions, in my view, would be the following: (1)

While technology in itself is a good thing, the **uses** to which it may be put is an entirely different matter over which even the technologically innocent may have useful contributions to make; (2) Globalization implies and is impossible without interaction and dialogue between the different components of the globe; (3) Dialogue further implies both talking and listening between the various dialoguing groups, each with its own peculiar modes, manner and style of expression; (4) In any genuine dialogue there must be the readiness and willingness on all sides, not only to **teach** but also to **learn**, even from the most taciturn of interlocutors (Wiredu, 1997, p.41).

CONCLUSION

In typical African fashion and manner, let me end with a true story which may also be a significant metaphor. As an undergraduate philosophy student, I once got an over-the-long-vacation assignment to 'describe God in the conception of your own people and then critically assess such description in the light of theism'. On holidays back home, I went to seek help from an elderly traditional Nso' priest. In the course of chatting about several other unrelated issues, I asked him: 'Taa (elder), how would you describe God, the way we, Nso' people see him?' He paused briefly and responded:

'You want me to describe God?'
'Yes', I confirmed eagerly.
'Are you a fool?', he queried.

I did not answer his rhetorical question, but carefully explained about my assignment. He said nothing for a long time and soon other visitors came in and the conversation shifted to other things. I thought he had forgotten about my question by the time I greeted him and started leaving, disappointment and frustration vaguely forming in my mind. But, just as I was about to step out of the door, he told me:'When you get back to school, ask your teacher for me whether he has nothing useful to teach you. He says you should describe God that where have you seen God? If you describe God correctly or wrongly, how would he know when he himself has never seen God?'

I went away in silence. (Silence is very important in traditional African informal pedagogy and learning.).

It is only recently that this encounter re-occurred to me when I started reflecting on the use we make of metaphors and how, very often, we forget that they are metaphors and take them quite literally, and how this can be the cause of a lot of trouble. 'Jesus is the son of God' or 'Mary conceived him without losing her virginity' are clearly metaphors. It even appears to me that our conception of God cannot but be a metaphor which may run us into many problems if we take it too literally. Faithful adherents of all the great Western religions would, no doubt, consider these views heretical. But it really appears to me that there may be some correlation between people's conception of God and their general attitudes and behavior. May it not be the case that people, groups, communities or cultures whose metaphor about God is too clear, definite, confident and uncompromising behave in like manner, which cannot but lead to fanatical extremism and intolerance when up against a similar but opposite attitude and behavior from others? Consider for a moment the intractability of the perennial disputes and conflicts within and between adherents of the different great Western faiths which have given birth to the worst forms of terrorism in the world today.

May it not be a matter of significance that all African religions are somewhat vague, non-doctrinaire, non-proselytizing, unsystematised and unorganized? African peoples and cultures also happen to be one of the most tolerant and accommodating on earth, with the broadest *live and let live* attitude and latitude.

It could be said that Western society today is one in which God is, more or less, dead, in spite of its monumental churches, cathedrals, mosques, synagogues, etc., relics of, perhaps, a more genuinely religious epoch, which now serve mainly the function of tourist attractions. But even though God may be dead in Western society, his attributes are evidently not dead but have been appropriated by Westerners and their science and technology, their economics and philosophies, their ethics, customs and laws. Today it is much easier to defy God than to defy the Western world. One remarkable thing about Western discourse in all domains of enquiry is the relative absence of **tentativeness**, especially in discussing issues and expressing views which are likely to be controversial because others are known to hold contrary views or opinions. It is not by accident that traditional Africans generally prefer to tell a proverb, parable, simile or metaphor in answer to a very difficult or controversial issue or question. The advantage here is that, since these figures of speech and literary devices are interpretatively inexhaustible, they can easily accommodate a variety of contrary or incompatible elements. The following questions should thus be posed: Is the Western conception and description of God as omniscient, omnipotent, omnipresent and omnibenevolent a disguised project description of Western epistemological ambitions, technological power, ubiquity and paternalistic philanthropy? Will the process of globalization be more than another school, for the rest of us, where Western doctrinal catechetical lessons are dispensed and imbibed?

REFERENCES

Potter, V.R. (1996): 'What does Bioethics Mean?' In *The AG Bioethics Forum*, Vol. 8, No. 1, pp. 2-3.

Tangwa, G. B. (1996): 'Bioethics: An African Perspective'. In *BIOETHICS*, Vol. 10, No. 3, pp. 183-200.

Wiredu, K. (1995): *Conceptual Decolonization in African Philosophy*. Ibadan: Nigeria.

----------- (1997): 'African Philosophy and Inter-Cultural Dialogue'. In *QUEST: Philosophical Discussions*, Vol. XI, No. 1-2, pp. 29-41.

CHAPTER FIFTEEN

LIVING IN A WORLD OF DIVERSITY AND VARIETY

[This chapter is developed from a presentation delivered on 21 October 2008, at the German Presidency, Schloss Bellevue, on the invitation of the German President, Horst Kohler, to a special intercultural group discussion on the general theme: „Forms of Modernity – Views of Modernity" and the particular topic: „Wisdom and the Pursuit of Knowledge"]

Wisdom is scattered in tiny little bits all over the world, amongst all peoples and cultures of the world

Introduction

The world in which we live (planet Earth) is marked by great variety and diversity. This diversity and variety can be perceived from every point of view in the various different peoples, cultures and languages of the world; in the various biological forms, both floral and faunal, that populate the earth; in the different ecological niches, climatic zones and geological forms of the world. Living creatures adapt themselves remarkably to the ecological niches in which they find themselves, merging themselves more or less harmoniously with the rest of nature.

At the very centre of the variegated variety and diversity of the world are human beings (conscious and thinking creatures), in their different races and cultures, speaking their different languages and trying to make sense of the world, to survive and to prosper in it. This has given rise, not surprisingly, to various conceptions and suppositions about the nature of human beings themselves, their relationship with the rest of nature and how they ought or ought not to live in the world. Human intervention in nature probably began with agriculture (about 10 millennia ago, we are told) when human beings could no longer survive, let alone prosper, by remaining mere hunters and wild fruit gatherers.

Western Culture and the Industrial Revolution

A watershed in the evolution of human interaction with and intervention in nature is marked by the European Industrial Revolution of the 18th and 19th centuries AD. The Western Industrial Revolution drew its impetus from the Baconian slogan that 'knowledge is power' convertible into commercial value, from the idea that all knowledge is unqualifiedly good, from the belief that nature is, in principle, completely knowable and controllable, and from perception of the universe as something which should be explored, subdued, dominated and exploited. Since the Industrial Revolution, technology has been propelled to great heights by commerce and the profit motive, by war and the will to dominate, by pure scientific curiosity and by the urge to improve human well-being. Today it can be said that technology is engaged in a life or death struggle with nature; a struggle liable to culminate in the death of nature, following the death of God, which can be argued to have occurred already, in the sense that new technologies within Western culture have effectively supplanted the place previously occupied by the old God of the Bible.

The Industrial Revolution and the technologies resulting from it greatly assisted Western imperial nations in their voyages of exploration, discovery, subjugation, colonization, domination and exploitation of other peoples. The cumulative effect of these very important achievements infused in Western culture an insatiable epistemological hunger, a spirit of omnivorous discovery, an automatic impulse to unifying, patenting, monopolizing and commercializing such discoveries and a penchant for spreading and promoting its ideas, vision, convictions and practices under the guise of universal imperatives of rationality and morality which ought to be binding on all and sundry. Today, Western culture is, indisputably, the dominant culture of the world – a descriptive fact, implying no value judgment. But Western culture is also a domineering, colonizing and proselytizing culture – a value judgment worthy of serious critical appraisal. In the domain of science and technology, Western culture is the acknowledged master at whose feet other cultures sit as pupils and apprentices, whether they like it or not.

But this does not mean that other cultures also have to learn and accept all other things from Western culture. Science and technology in themselves have nothing to do with how, for example, people consider their coexistence with other creatures of the earth, conceive of or worship God, how they marry or bring up their children, how they organize their social system or even what they think of the uses to which technology should be put.

Human Limitations and Myth-Making

Humans, no matter where and how they are situated in the world, are necessarily concerned about both the descriptive (is) and prescriptive (ought) aspects of being in the world. But human efforts both at describing (science) and prescribing (ethics) human being (existence) in the world, come up against what appear like ineradicable human limitations which need permanently to be factored into all human efforts at understanding, describing and prescribing the world. These limitations include human fallibility, vulnerability, imperfection, mortality. Given these limitations and in spite of remarkable human achievements in many domains, human ambitions need to be tentative, thoughtful and cautious.

Human limitations make human myth-making an inevitable part and parcel of human existence and attempt at explaining and living in the world. Simple questions like "Why are we here? How did we get here? How should we live here? And where are we going from here?" cannot be answered without resorting to myths and metaphors. All human cultures engage in myth-making although, as Michael Novak has remarked [*The Experience of Nothingness*, New York: Harper and Row, 1970, p.16], we may not be fully aware that our own lives and culture are shaped by myths. We may tend to look on myths as what people in 'those other cultures' believe in while within our own culture we deal only with 'reality'. But, just as no individual human being can perceive from more than a single point of view, no human culture can perceive reality in a comprehensive and holistic manner and none can do without myths. Such myths, whether they are of a religious or scientific kind, set the parameters of each people's and each culture's ambitions and realizations. The myths we live by are deeply significant of our convictions, our ambitions, and the possibilities of our future achievements. Here are two of such myths, representative of two living cultures of the modern world: the industrialized Western world and the traditional African world.

The Genesis Story

In the beginning God created the heaven and the earth. And the earth was without form, and void; and darkness was upon the face of the deep. And the Spirit of God moved upon the face of the waters. And God said, Let there be light: and there was light. And God saw the light, that it was good: and God divided the light from the darkness. And God called the light Day, and the darkness he called Night. And the evening and the morning were the first day. And God said, Let there be a firmament, and divided the waters which were under the firmament from the waters which were above the firmament: and it was so. And God called the firmament Heaven. And the evening and the morning were the second day. And God said, Let the waters under the heaven be gathered together unto one place, and let the dry land appear: and it was so. And God called the dry land Earth; and the gathering together of the waters called he seas: and God saw that it was good. And God said, Let the earth bring forth vegetation, the plants yielding seed, and the fruit tree yielding fruit after its kind, whose seed is in itself, upon the earth: and it was so. And the earth brought forth vegetation, and plants yielding seed after their kind, and the tree yielding fruit, whose seed was in itself, after its kind: and God saw that it was good. And the evening and the morning were the third day. And God said, Let there be lights in the firmament of the heaven to divide the day from the night; and let them be for signs, and for seasons, and for days, and years: And let them be for lights in the firmament of the heaven to give light upon the earth: and it was so. And God made two great lights; the greater light to rule the day, and the lesser light to rule the night: he made the stars also. And God set them in the firmament of the heaven to give light upon the earth, And to rule over the day and over the night, and to divide the light from the darkness: and God saw that it was good. And the evening and the morning were the fourth day. And God said, Let the waters bring forth abundantly the moving creature that has life, and fowl that may fly above the earth in the open firmament of heaven. And God created great sea creatures, and every living thing that moves, which the waters brought forth abundantly, after their kind, and every winged fowl after its kind: and God saw that it was good. And God blessed them, saying, Be fruitful, and multiply, and fill the waters in the seas, and let fowl multiply in the earth. And the evening and the morning were the fifth day. And God said, Let the earth bring forth living creatures after their kinds, cattle, and creeping things, and beasts of the earth after their kinds: and it was so. And God made the beasts of the earth after their kinds, and cattle after their kinds, and everything that creeps upon the earth after its kind: and God saw that it was good. And God said, Let us make man in our image, after our likeness: and let them have dominion over the fish of the sea, and over the fowl of the air, and over the cattle, and over the earth, and over every creeping thing that creeps upon the earth. So God created man in his own image, in the image of God created he him; male and female created he them. And God blessed them, and God said unto them, Be fruitful, and multiply, and fill the earth, and subdue it: and have dominion over the fish of the sea, and over the fowl of the air, and over every living thing that moves upon the earth. And God said, Behold, I have given you every plant bearing seed, which is upon the face of all the earth, and every tree, *which has seed in its fruit; to you it shall be for food. And to every beast of the earth, and to every fowl of the air, and to everything that creeps upon the earth, wherein there is life, I have given every green plant for food: and it was so. And God saw everything that he had made, and, behold, it was very good. And the evening and the morning were the sixth day.*

Thus the heavens and the earth were finished, and all the host of them. And on the seventh day God ended his work which he had made; and he rested on the seventh day from all his work which he had made. And God blessed the seventh day, and sanctified it: because in it he had rested from all his work which God created and made. [Genesis, Chapter 1: 1-31; Chapter 2: 1-4; The King James 2000 Version].

The Calabash of Wisdom
Ijapa was very strong, very intelligent and clever, but also very proud, cunning and greedy. One day, while God was deeply asleep after having drunk too much palm wine, Ijapa sneaked into heaven and stole the calabash of wisdom from the foot of God's bed. He tied the calabash around his neck with a rope so that it was hanging below his chest and ran away. As he was making his way home he discovered that a big tree had fallen across his path. Three times he tried to climb over the fallen tree and three times he failed. Meanwhile a little bird perched high on the branch of another tree had been watching him. The bird laughed and addressed him: "What a fool you are! Don't you realize that it is that big calabash hung around your neck that is preventing you from crossing that log? If you take it off your chest and put it instead on your back or shoulder, you will cross that log easily." Ijapa was so ashamed to realize his own stupidity and so humiliated at being so disgraced by a small bird that, in anger, he took off the calabash of wisdom from his neck and smashed it on the tree trunk and it shattered into little bits. From that day, wisdom is scattered all over the world in tiny little bits.

Conclusions
It is evident in the first myth that God is a being whose chief attributes are omnipotence, omniscience, and omni-benevolence, a being rather accurately mirrored by Western culture in its main accomplishments and ambitions. In the second myth, God's attributes are difficult to deduce, but s/he is evidently neither omnipotent nor omniscient, from the fact of getting drunk on palm wine, lazily sleeping like a drunkard and losing the all important calabash of wisdom to a clever human thief - again a rather telling mirroring of African culture and Africans, lazily sleeping besides

enormous wealth as all manner of thieves cart it away. The morale of the first myth is exemplified, par excellence, in the technological revolution culminating in current biotechnologies and their future potentialities. The morale of the second myth draws attention to human limitations, imperfections, fallibility and consequent need for procedural caution and circumspection.

Experience may not be the only source of human knowledge but it certainly is the most important. Human experience not only is inevitable but it also inevitably counts. And one of the most important ways human experience (knowledge) counts is in the **uses** to which it is put. We need to reverse our guideline from "Wisdom and the pursuit of knowledge" to "Knowledge and the pursuit of wisdom". It is not knowledge we need to pursue; indeed we have more than enough knowledge at our disposal. What is lacking, what is critically at stake, is using available knowledge in wise ways capable of enhancing human welfare and well-being in the world.

Human experience and knowledge are too vast and too complex for any one person, group of persons or culture. What is true in this regard of individual human beings, namely, that, no matter how knowledgeable and self-sufficient you may be, any other human being, taken at random, no matter how ignorant and helpless s/he may appear, knows at least something that you do not know, is even truer of human cultures in general. Modernization, no matter amongst which peoples or cultures, is inconceivable today without use of Western science and technologies. But it earnestly is to be hoped that Western science and technologies can be put to more humane and wise uses, taking into more serious consideration the dignity and moral equality of all human beings, irrespective of their widely differing existential situations and material conditions, than has hitherto been the case. The world in which we live today has less need of further scientific discoveries and continuous technological innovations than it has need of ventilating and disseminating the fruits of already made discoveries and innovations to all and sundry. Enormous resources for enhancing human welfare and well-being exist in the world today. It is a great scandal of our time, which calls into question human rationality, that millions of human beings daily die of hunger, poverty and disease.